摩登样板间 IV
欧式风尚

Modern Show Flat IV
European Fashion

ID Book 图书工作室 编

中国·武汉

A BRIEF TALK ABOUT SHOW APARTMENT DESIGN

Show apartment is an ideal home space created by property developer to attract target customers. The designer focuses on creating a spectacular visual effect, emphasizing on expressing the characteristics of real estate and the target customers' ideal life.

In the process of determining the design theme, we will first communicate with the real estate agency to know more about the real estate's characteristics, the local cultures background and the needs of the target customers. And then we focus on the case's price as well as the community the developer wants to sell. After that we mine the requirements and interest of the target customers, and think about the impressive scenes that may be move them, etc..

The design process's order is building exterior, interior architecture and spatial design. The first two parts are finished by the architects, the latter part is completed by an interior designer. But the architect does not have the experience on space design, which would lead to unreasonable spatial structure of the layout. Therefore, at the beginning of the design, we will get into the design of the building, making the building surface and spatial structure more reasonable, after which we are more handy in the interior design.

Comparing the facade with the plane, we pay more attention to the plane. We think the plane is more important than the facade, because we have think about the latter when we do the plane. Therefore, when we pull up the facade it is really a very good work. We often use "gray space" approach, and use the transitional space between the interior and external environment to achieve the indoor and outdoor integration. We use "gray space" to increase the level of space, coordinating different functional building to be perfect. We change the space ratio to make up the architectural layout design and enrich the interior space.

All in all, we have three principles on the design of show apartment. Firstly, use the advantages. Secondly, avoid the shortness. Houses always have defect, what we should do is to find a way to make up the shortcomings. Thirdly, the theme; a show apartment must has a clear idea or theme style, so that people can have a deep memory.

DaE Design / Raynon Chiu Designer Studio (Taiwan)
Qiu Chun Rui

浅谈样板间设计

　　样板间是地产开发商为吸引目标客户而精心打造出来的理想家居空间。设计师的设计重点在于营造引人入胜的视觉效果，注重表现楼盘的特质及展现目标客户理想的生活品位。

　　在确定设计主题的过程中，我们首先会与房产商沟通，了解楼盘的特质、当地的文化背景及目标客户群的需求。关注案子本身的售价，以及开发商想要销售的群体。挖掘这一类目标客户群的使用要求及兴趣爱好，并思考可能会打动他们的生活场景等。

　　设计的先后次序分别是建筑外观、内部结构、空间设计，前两部分由建筑师完成，后一部分则由室内设计师完成。但建筑师并非就具备空间设计的经验，这会导致空间结构布局的不合理，因此在设计之初我们也会介入建筑的设计，使建筑外观与空间结构布局都更加合理。使得我们在之后的室内设计中更加得心应手。

　　相比而言，我们更注重平面，认为平面比立面要重要，因为我们在做平面的时候就把立面考虑到了，这样当我们把立面拉起来后就已经是很好的作品了。我们常使用"灰空间"的手法。利用室内与其外部环境之间的过渡空间，来达到室内、外融和的目的。用"灰空间"来增加空间的层次，协调不同功能的建筑单体，使其完美统一。改变空间的比例，弥补建筑户型设计的不足，丰富室内空间。

　　总的来看，我们对于样板间的设计有三个原则。扬长，充分展示自己的优点。避短，通过设计的手法来弥补户型的缺憾，房子或多或少都存在某些缺憾，需要通过设计给予弥补或掩饰。主题风格，样板间设计必须有明显的主题思路或风格，让人们记忆深刻。

台湾大易国际设计事业有限公司•邱春瑞设计师事务所

邱春瑞

目录

- 006　源于戈尔德的法式乡村
- 012　熙璟城豪苑样板房
- 020　玫瑰小筑独幢别墅
- 028　荣禾·曲池东岸二期E户型
- 034　融汇桂湖样板房
- 040　深圳祥祺·滨河名苑样板房
- 044　玩家样板间
- 050　都市新贵·南屿翡丽湾样板间
- 054　A公寓
- 064　陌上花开
- 068　昆明中豪悦城花园样板房
- 076　有色生活
- 086　繁华之上·ECO城样板房
- 092　武汉天地
- 098　中央公园住宅
- 104　静和之美
- 110　世茂臻园
- 116　海航城E户型样板房
- 124　丰泽园一期
- 134　深圳海航城C2户型样板房
- 140　雅奢素影
- 150　蓝羽雀
- 156　白色恋歌
- 162　和昌森林湖D1-8户型样板房
- 168　融侨城某宅
- 174　凤翔新城某宅
- 180　宅心物语
- 186　尖沙咀名铸
- 190　保利赛纳维拉某样板间
- 196　福州融信大卫城别墅
- 206　顺发康庄某宅
- 214　深圳曦城别墅
- 224　古典美学的华丽转身
- 232　天鹅湾某住宅
- 240　奥克斯盛世缔壹城中央样板房
- 248　情归于家
- 252　保利塘祁路A户型样板房
- 256　棠梨天城
- 264　雅致简欧
- 268　融侨观邸某宅
- 276　中海万锦江城某样板间
- 282　眉山凯旋国际广场样板间
- 286　红磡海逸豪园
- 290　漫步人生
- 298　和昌森林湖G2-6户型样板房
- 304　和昌森林湖G2-5户型样板房
- 308　恬静心居
- 314　清韵昂然

Contents

006	French countryside originated Gordes
012	Xi Jing City Haoyuan Show Apartment
020	Rose House Single Villa
028	Ronghe•Quchi East Bank Phase II E Unit
034	Rong Hui Gui Lake Show Apartment
040	Shenzhen Xiangqi•Binhe Mingyuan Show Flat
044	Player Show Apartment
050	Modern New Noble•Nan Yu Fei Li Bay Show Flatt
054	A Apartment
064	Mountain Blossom
068	Kunming Zhonghao Yue City Garden Show Flat
076	Colorful Life
086	Over the Busy•ECO City Show Apartment
092	Wuhan Space
098	Central Park House
104	The Beauty of Peace
110	Shi Mao Zhen Yuan
116	Hai Hang City E Unit Show Apartment
124	Feng Ze Yuan First Phase
134	Show Flat for C2 House Type, Shenzhen HNA City
140	Ya She Su Ying
150	Blue Finch
156	White Love Song
162	Hechang Forest Lake D1-8 Unit Show Apartment
168	Rong Qiao City X House
174	Feng Xiang New City X House
180	Homey Heart Story
186	Tsim Sha Tsui Ming Zhu
190	Baoli Saina Vera X Show Apartment
196	Fuzhou Rongxin Dawei City Villa
206	Shunfa Kangzhuang X House
214	Shenzhen Xi City Villa
224	Splendid Turning of Classical Aesthetics
232	Swan Bay X House
240	AUX Golden First City Central Show Apartment
248	Charm in Home
252	Baoli Tangqi Road A Unit Show Apartment
256	Begonia and Pear Flower City
264	Elegant Simple European
268	Rong Qiao Apartment X House
276	Zhonghai Wanjin River City X Show Apartment
282	Meishan Triumphant International Square Show Flat
286	Hunghom Harbour Plaza Garden
290	Stroll On The Way of Life
298	Hechang Forest Lake G2-6 Unit Show Apartment
304	Hechang Forest Lake G2-5 Unit Show Apartment
308	Quite Heart Space
314	Full of Charm

摩登样板间 IV
欧式风尚

源于戈尔德的法式乡村

French countryside originated Gordes

设计单位：SCD（香港）郑树芬设计事务所

设计师：郑树芬

参与设计：杜恒、陈洁纯

项目面积：139 m²

项目地点：广东深圳

Design Company: SCD (Hong Kong) Simon Chong Design firm

Designer: Simon Chong

Participated Designer: Amy Du, Holiday Chen

Project Area: 139 m²

Project Location: Shenzhen Guangdong

MODERN SHOW FLAT IV **EUROPEAN FASHION**

本案最具特色之处，就是精致的设计细节与天然材料的和谐融合，呈现出一个舒适自然的居住环境。客厅以木褐色为主色调，线织纹理的磨砂壁纸覆裹着客厅的墙壁。青木色的具有厚麻布质感的窗帘、人字型实木复合地板，勾勒出天然雅朴的空间气质。富有光泽的牛皮棕色沙发、米色粗麻布面沙发与橄榄绿布面提花单人沙发相组合，造就了质感丰富的混搭效果。明媚多彩的花簇将法国南部旖旎的乡村气息引入空间，点亮了空间的色调，使之由内而外地散发法国乡村风情的气质。

餐厅摆放着不经雕琢的木制餐桌，造型简单、线条直接，原生态的纹理烙印着岁月流转的痕迹，木质餐椅则增加了绸面提花的细节，同种材质的原生态木柜在各个功能空间也能觅得踪影。

主卧的每个细节都值得细细品味，流行于路易时期的经典法式印花棉布出现在床品、床头软包、床尾凳上，地毯的图案及蓝青色调与此相呼应；墙面上的大幅挂画恬淡轻柔，符合卧室温馨浪漫的气质，摆饰、细节上的贯通让整个主卧饱满而和谐。

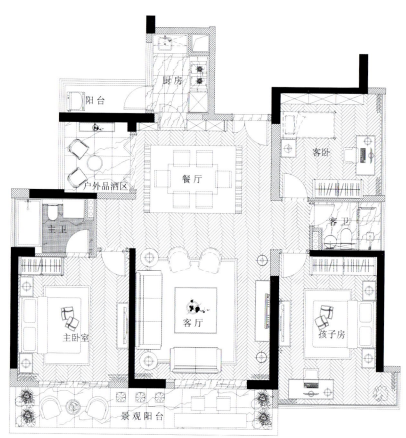

淡黄色的条纹壁纸给儿童房带来活力与朝气，运动元素的融入塑造了小男孩活泼好动的鲜明个性。色调清淡的花卉床品为次卧带来清新之感，加入柠檬黄的点缀，让空间更加明朗。浴室以白色、淡黄色色调为主，质朴大方，马赛克的运用则增强了法式乡村的韵味。

The most unique way, this case is delicate design details with the harmony of the natural material, this case is delicate design details and harmonious fusion of natural material, presents a natural and comfortable living environment. The sitting room is given priority to with wood brown color, line of weave texture frosted wallpaper wrapped in the wall of the sitting room; Aoki color with thick linen texture of curtain, herringbone solid mu fu joins a floor board, draw the outline of the space of natural park, temperament. Shiny brown leather sofa, the sofa beige burlap surface combined with olive green cloth jacquard single person sofa, make the rich texture of mixing effect. Bright and colorful flowers to the south of France the charming rustic into space, lit up the room is tonal, make it by sending out the temperament of French countryside amorous feelings from the inside out.

Restaurant with wooden table without carve, modelling is simple, line is direct, the texture of the original branded the marks of years flow, wooden chair increased silken face details of jacquard, the same material of raw wood cabinet in each functional space also can find nowhere to be seen.

Advocate lie to be worth to savor every detail, popular in the classic French Louis period cotton print in bed is tasted, soft package of the head of a bed, bed end stool, the pattern of carpet and indigo color photograph echo with the; Sharp hang a picture on the wall light soft, conform to the bedroom sweet and romantic temperament, place adorn, details on the breakthrough let whole advocate lie full and harmonious.

Pale yellow stripe wallpapers bring vitality and vigor to children room, sports elements into lively personality shaped the little boy. Color light flower bed is tasted for second lie bring pure and fresh feeling, add lemon yellow ornament, let a space more clear. The bathroom is given priority to with white, light yellow color, simple and easy, the use of Mosaic will be improved through the French countryside flavor.

摩登样板间 IV
欧式风尚

熙璟城豪苑样板房

Xi Jing City Haoyuan Show Apartment

设计单位：深圳创域设计有限公司、殷艳明设计顾问有限公司

项目面积：140 ㎡

项目地点：广东深圳

主要材料：橡木、墙布、茶色不锈钢、灰茶镜、皮革、混拼马赛克

Design Company: Shenzhen Chuangyu Design Co., Ltd., Yin Yanming Design Consultation Co., Ltd.

Project Area: 140 ㎡

Project Location: Shenzhen Guangdong

Major Materials: Oak, Wall covering, Brown stainless steel, Gray tea mirrors, Leather, Mixed fight mosaic

奢于内，形于外

熙璟城公共区域的入户大堂是用星级酒店的理念来打造的，充分利用空间层层递进的关系，营造了具有时尚气息的序列空间感，对于空间的整合和优化可圈可点，进一步提升了小区整体的公共空间档次。

本案整体设计布局清晰、流线顺畅，以米色、中灰色、深褐色为主调呈现的空间氛围突现出简欧的华丽，地面铺装立体几何形体的运用又融合了时尚现代的元素，中式方格铁艺与仿云石灯片的组合以混搭手法诠释了空间的文化感，局部东方风格元素与欧式手法的碰撞使空间的意象有了更深层次的表达，也充分表露出设计师成熟而大胆的设计手法运用，设计手法只要运用得当，一切都是自然而有趣味的，也因此让空间充满了另一种视觉意象美。

灯具设计也是现代和欧式不同手法的组合，灯光运用合理，贝壳马赛克的点缀和大型植物组合油画给空间增添了人文情怀，雕塑的点缀给予区域空间更多的想象和魅力。

在空间演绎大堂的气派，总有些亮点是让人记忆犹新的，电梯厅墙面的光、毛面石材蚀刻的"花"形图案就是如此，它是独有的，让客人候梯时总分有那么一瞬能够会心地一笑，设计如此，已足够。

Inner luxury, outer beauty

Xi Jing City Public Area's lobby is created under the concept of star hotel. The building uses spatial gradation to create fashionable spatial sense. The integration and optimization promote the public house's grade.

The overall design of the layout is clear and smooth, using beige, gray, dark brown as tone color showing simple-European's gorgeous sense. The floor three-dimensional geometric shape has modern elements. The combination of Chinese grid iron and imitation marble

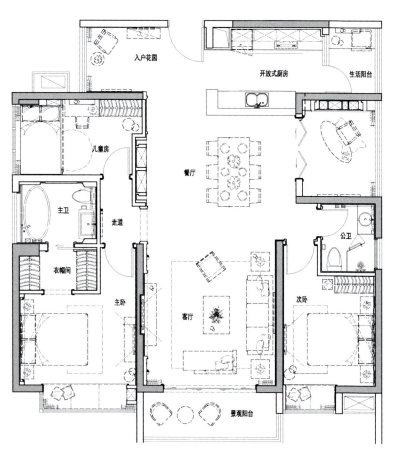

interprets the culture sense of the space. The collision of local oriental style elements with European techniques makes space imagery gains a deeper expression, and also fully reveals mature and bold design techniques. Since the design technique is used properly, everything is natural and interesting, and the space is filled with visual imagery beauty.

Lighting design is another combination of modern and European approach. Lighting is used reasonably, shell mosaic decorations and large plants painting add humanistic feelings to the space; sculpture ornament gives more imagination and charm to the whole space.

The interpretation of the lobby style is unforgettable. The light of the TV room and "flower" pattern of the stone material are unique. The design aims to make the guest smile during the waiting time in the elevator.

摩登样板间 IV
欧式风尚

玫瑰小筑独幢别墅

Rose House Single Villa

设计单位：深圳市派尚环境艺术设计有限公司

设计师：李益中、沙千帆

项目面积：325 m²

主要材料：拼花木地板、印尼酸枝木、灰木纹大理石、刻花大理石、壁纸、地毯

Design Company: Shenzhen Paishang Environment Art Design Co., Ltd.

Designer: Li Yizhong, Sha Qianfan

Project Area: 325 m²

Major Materials: Parquet, Indonesian rosewood, Ash wood marble, Carved marble, Wallpaper, Carpet

别墅位于安徽芜湖凤鸣湖天然水域，悠然的凤鸣湖水静静流淌，周围芳草萋萋、绿树掩影，别有一番欧式小镇自然闲适的风情。设计师采用了美式风格来装饰，它大都带给人大气、沉稳，舒适的感觉，但稍加点缀变换，也能营造出古典时尚的感觉。美式风格的宽大、舒适，用在这栋独幢别墅里，显得恰到好处，也给了设计师很大的发挥空间。

在空间布局上，表现极为突出的是起居室、玄关与餐厅之间，设计师将起居室与餐厅布置成一条直线，然后与玄关成90°直角，如此，便增强了起居室和餐厅以及玄关之间的流通性，让空间更加灵动，视野也更加开阔。在平面布置上，原有的布局比较合理，基本上不需要做太多的改动，所以，留给设计师发挥的便是各种材料、家具、灯具、配饰的选择与搭配。即如何在空间内打造出一种既时尚大气，又兼具奢华感觉的美式风情。

起居室里，深色的拼花木地板，墙面画框里的鹿头木头挂饰，古朴的铁艺吊灯和壁灯，呈现出美

式惯有的粗犷，自然与随意。香槟金色的窗帘、银色的脚踏，又在美式的大气中增加了华贵的气息。白色的顶棚、门套、踢脚线，还有白色的纱帘，纯白的绣球花点缀其中，产生了一种古典而纯净的美。

自然、奢华、古典的气质贯穿在整个居室设计中。二楼主卧中黑白相间的床品、白色的百叶窗、红色的装饰品，色调稳重且具有品位。以红色为主调的女孩房，配以白色的床品、白色的台灯和落地灯，时尚又有着公主般梦幻的气质。在设计中，设计师巧妙地运用色彩的搭配、材质的选择、配饰的点缀，华美的感觉油然而生。

Villa is located in Fengming Lake natural waters in Wuhu, Anhui, with quietly flowing Fengming water. It is surrounded by lush grass and trees with a European style town natural leisure. The designer uses American style to decorate the space; it brings people calm and comfortable feeling. A slight change can also create a classic fashion sense. American style's large, comfortable looks just right in this villa and also gives the designer more development space.

On the spatial distribution, it is worth to discuss the space in the center of the bedroom, hallway and the dining room. The designer arranges the bedroom and the dining room into a straight line, forming 90°right angle with the hallway. Therefore, it enhances the liquidity of the bedroom and dining room as well as the hallway, making the space more vivid, the vision wider. On the layout, the original layout is reasonable, basically do not need to do too much change, however the designer can do different on varieties materials, such as furniture, lamps and selection of accessories. He need to think about how to make a fashion, bold and luxury American feeling.

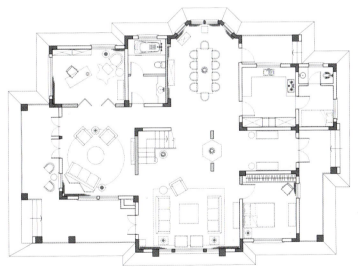
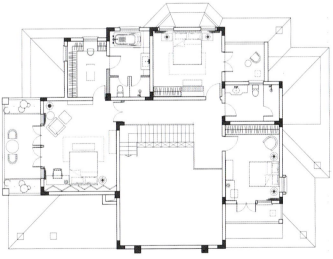

In the living room, the dark parquet flooring, the deer wood ornaments in the frame, the ancient wrought iron chandeliers and wall lamps reflects the rough American style, naturally and casually. Champagne gold curtains and silver pedal add luxurious atmosphere in the space. White roof, door pockets, foot line, and white gauze and white hydrangea, produce a classical and pure beauty.

Natural, luxurious, classical temperament is throughout the entire room design. Black and white bedding in the main bedroom on the second floor, white shutters and red decorations are stable and stylish. Red tone girl house with white bedding, white table lamps and floor lamps are stylish and has a princess-like fantasy temperament. In the design, the designer cleverly uses color, materials, accessories to create a beautiful feeling.

摩登样板间 IV
欧式风尚

荣禾·曲池东岸二期E户型

Ronghe•Quchi East Bank Phase II E Unit

设计单位：SCD（香港）郑树芬设计事务所

设计师：郑树芬

参与设计：杜恒、丁静

项目面积：195 m²

Design Company: SCD (Hong Kong) Simon Chong Design firm

Designer: Simon Chong

Participatory Designers: Amy Du, Jing Ding

Project Area: 195 m²

本案的整个空间精致闪亮却不奢靡，充满着诗意的质感，与施华洛世奇水晶高贵、优雅的美丽不谋而合。而施华洛世奇水晶的标志——天鹅，则象征着纯洁，代表着力量，连接着所有向往美好与快乐的心灵，这也正是设计师对该空间的美好寓意——人与人之间心灵沟通的家园。

浅色调的绒布沙发既典雅又现代，线条简约的施华洛世奇水晶吊灯为客厅带来轻奢感，地毯和挂画以简单抽象画为主，大气的笔法和泼墨的轻松和自由感，正体出现代都市人的向往。印有克里姆特经典作品《吻》的陶瓷装饰，唯美的拥吻画面昭示着温馨、浪漫和富有激情的生命力，充满了爱的力量。更是与施华洛世奇的纯洁之美相得益彰，升华客厅空间的感染力。

蓝紫色圆形地毯、湖蓝色圆形蒲公英图案餐椅，搭配璀璨的水晶吊灯，塑造出愉悦、轻松的就餐氛围。香芋紫色的三人绒布沙发与浅棕方格布面沙发混搭，深棕色的窗帘则延伸了空间的层次感，简洁白色立柜、金色造型摆设与黑白几何抽象画渗透出现代都市气息。

主卧犹如走进浪漫的色彩世界，柠檬黄色的床品点亮了整个米色空间，丝绒墨蓝色窗帘，嫩粉色桃花，欧式花纹壁纸等，都令人仿佛置身于花花世界，感受到春天般的芳香与温暖，充满生机与力量。

The case's whole space is fine, shining but not extravagant, which is filled with poetic texture, and Swarovski crystal's noble, elegant beauty. The Swarovski crystal logo - Swan, is a symbol of purity, representing strength, connecting all the soul longing for happiness, which is the beautiful meaning the designer giving the space - a heart harbor between people.

Light color velvet sofa is elegant and modern, simple lines of Swarovski crystal chandeliers light bring a sense of luxury for the living room. Carpets and paintings are mainly simple abstract style. Relaxed atmosphere and free sense of ink strokes show the modern city people's yearning. Pottery decorations printed with Klimt's famous "The Kiss" shows the warm, romantic and passionate vitality, and full of the power of love. The picture is echoing with Swarovski's purity, enhancing living room's appeal.

Blue-purple circular rug, lake blue circular pattern dandelion chair, and a dazzling crystal chandelier create a pleasant, relaxed dining atmosphere. Purple velvet sofa and three-people brown sofa are mixed with each other. Dark brown curtains extend the space's layers, simple white wardrobe, gold ornaments and black and white geometric abstract shapes permeate the modern urban atmosphere.

The master bedroom is like a romantic colorful world, lemon yellow bedding lights up the whole beige space. Ink blue velvet curtains, soft pink flower and continental pattern wallpaper are making people living into a colorful world. The spring-like fragrant and warm feeling is full of vitality and strength.

Rong Hui Gui Lake Show Apartment

设计单位：深圳市昊泽空间设计有限公司

设计师：韩松

项目面积：100 ㎡

主要材料：帕拉米黄大理石、水洗白木饰面、胡桃木木地板、蓝色壁纸

Design Company: ShenZhen Horizon Space Design Co., Ltd.

Designer: Han Song

Project Area: 100 ㎡

Major Materials: Parami Yellow Marble, Unvarnished wood veneer, Walnut wood floors, Blue wallpaper

海的声音

地中海风格以其极具亲和力的色彩及装饰组合搭配上的大气被人们所喜爱。蔚蓝色的海岸和白色沙滩，是人们对"面朝大海，春暖花开"生活情景的一种向往。纯美的色彩组合，蓝白色搭配的色调，成就了地中海风格家居的无限魅力。

设计师以地中海风格完美演绎了整个空间，造型以简洁的直线和柔美的曲线为主，用白色的乳胶漆、蓝色的马赛克、简洁的铁艺装饰和蓝色条纹的家具作为装饰，打造出一个简约时尚、清爽宜人的家居环境。室内空间采用大面积的蓝、白色，诠释着人们对蓝天白云、碧海银沙的无尽渴望。以代表海洋的蔚蓝色为基底色调贯穿整个空间，使之充满自然、浪漫的生活气息。富有诗意的黑色铁艺、具有时代感的洗白木色、亚麻质地的家居，似乎都被这清新的色调包裹着，相互融合在一起。无论在空间的任何一个角落，都能体会到业主悠然自得的生活态度和阳光般明媚的心情。

Sound of the Sea

Mediterranean style is loved by the people with its great affinity and the combination of colors and decorative atmosphere. Azure coast and white sand beaches is what people longing for as an ideal life situation of "spring blossoms". Pure color combinations, blue and white match are the infinite charm of the Mediterranean-style home.

Designer uses a Mediterranean style perfectly interpreting the entire space. The design bases on simple lines and beautiful curves, combining with white latex paint, blue mosaic, simple decorative wrought iron decoration and blue stripes furniture, creating a simple and stylish, refreshing and pleasant home environment. Indoor space uses large area of blue, white, showing people's desire for the endless blue sky. Designer uses ocean blue as a tone throughout the whole space, making the space full of natural, romantic life atmosphere. The poetic black wrought iron, white-washed wood color, linen texture household, all seems have been wrapped in the fresh colors, merging together. No matter any corner in space, the owners can experience the leisurely attitude towards life and the bright and sunny mood.

摩登样板间 IV
欧式风尚

深圳祥祺·滨河名苑样板房

Shenzhen Xiangqi • Binhe Mingyuan Show Flat

设计单位：深圳市盘石室内设计有限公司 / 吴文粒设计事务所
陈设公司：深圳市蒲草陈设艺术设计有限公司
设计师：吴文粒、陆伟英
参与设计：陈东成
开发商：深圳祥祺地产
主要材料：索菲特金大理石、流金啡大理石、帝皇金大理石

Design Company: Shenzhen Huge Rock Interior Design Co., Ltd. / Wu Wen Li Design Form
Display Company: Shenzhen Pucao Furnishings Design Co.,Ltd
Designers: Wu Wenli, Lu Weiying
Participatory Designer: Chen Dongcheng
Developer: Shen Zhen Xiangqi Real Estate
Major Materials: Sofitel gold marble, Golden brown marble, Emperor of marble

邂逅巴黎之意

室内空间中处处流露着一种高品质的生活语言，薄而淡的金色情怀，更显示出人生的阅历和生命的厚度，百"阅"不厌、韵味十足。进入客厅，仿佛来到巴黎街头的咖啡馆，几分高雅、几分浪漫、几分沉稳的氛围颇具感染力。妩媚波澜的印花绸缎，让人回味起那开在巴黎街边的金百合，婀娜多姿，博得家人嫣然一笑。巴黎的浪漫，贵在具有生机盎然的气息。生命在本我中成长，是一种品质、一种追求，使得人能在沧海横流中站稳脚跟，在浮名虚化中找准自己的位置。

轻奢华的奢华是一种优雅中的奢华，"舒适、品质、享受"才是它的目的，而非为了炫耀。当你进入此空间中，就能感受到空间之间精致且舒适的微妙氛围。摒弃了欧式元素的符号化、模式化，而将传统繁杂的装饰细节融入到诸如平面构成、疏密对比、色彩对比这样的现代设计思想中，在视觉上让人更加舒适、悦心。在这高雅、温馨、浪漫的空间里自然而然的让人们静下心来思考自己的工作、爱情与生活，触及内心最细腻的心弦。

Encounter Paris

Indoor space shows a high quality life, thin and pale golden feelings reveals the thickness of life. The space is full of flavor, people will not feel boring in it. Getting into the living room, as if get on the street of Paris' cafes, with somewhat elegant, somewhat romantic, somewhat calm. Print silk waves make people think about the charming and graceful Paris Golden Lily, winning family smiling. Paris is romantic, with

a vitality in your breath. The growth of life is a quality, a pursuit, so that people can gain a firm foothold in the sea flow while find their position in the vain world.

Light luxurious luxury is an elegant luxury whose purpose is "comfort, quality, enjoyment" but not for flaunt. When people stepping into the space he or she can feel the elegant and delicate and subtle atmosphere between the spaces. It abandons the symbolic mode of European elements, while adds the traditional decorative details into the complex plane, such as composition, density contrast, color contrast and other modern design ideal. Visually it makes people feel more comfortable and happier. In this elegant, warm, romantic space, people naturally stop and think about their own work, love and life, touching the most delicate heartstrings.

摩登样板间 IV
欧式风尚

玩家样板间

Player Show Apartment

设计单位：多维设计
设计师：张晓莹
项目地点：四川成都
主要材料：木饰面板、镜钢、木地板、壁纸

Design Company: Duo Wei Design
Designer: Zhang Xiaoying
Project Area: Chengdu Sichuan
Major Materials: Wood panels, Mirror steel, Wood floors, Wallpaper

MODERN SHOW FLAT IV EUROPEAN FASHION 045

本案摒弃了繁复厚重的古典欧式，进行简化创造后，结合都市元素为年轻一代带来充满活力的现代简欧家居。新与旧、低调与奢华、优雅与随意的有机结合，碰撞出精彩的设计火花。由色块、几何图形、线条组合的装饰品点缀在空间各处，恰到好处地提升空间的现代设计感，各个细节凸显出独特的魅力，从局部一直蔓延到整体，赋予了空间更多的雅致与奢美，完美体现了设计师的设计主张。

空间里的每一处设计都是经过设计师深思熟虑的，大到家具、墙饰的挑选，小到杯具、抱枕的选择，无不透露出一种精致的生活状态。欧式风格的设计为空间注入非凡的精神象征。空间整体营造出一种和谐、温馨的氛围，伴随着设计师娴熟的手法，将爱的力量、生活的激情融汇于空间的每一处，演绎出别具一格的魅力家居。

The captioned case abandons the complicated and heavy classical European style. The newly simplified and creative style and the modern elements bring vibrant modern simple European home furniture to urban young generation. Old and new, low-key and luxury, casual and elegant combination collides a wonderful design sparks. The decorations made by colors, geometric shapes, lines dots around the space, just enhancing the sense of modern design in the space. Each detail highlights the unique charm from the local to the whole, giving the space more elegant and luxury beauty to the space, perfectly embodies the designer's design ideas.

All the design in the space are considered deliberately by the designer. No matter the selection of the furniture and wall hangings, or the small cups and pillows, all reveal a refined living conditions. European-style design injects extraordinary spiritual symbol into the space. The overall space creates a harmonious and warm atmosphere. The designer put the power of love and passion into every corner of the space along with skillful way, showing unique charm home furniture.

Modern New Noble • Nan Yu Fei Li Bay Show Flat

设计单位：深圳张起铭室内设计有限公司

设计师：张起铭

参与设计：詹远望、魏宗全

项目面积：82 ㎡

主要材料：石材、玫瑰金、墙布、木饰面

Design Company: Shenzhen Zhang Qiming Interior Design Co., Ltd.

Designer: Zhang Qiming

Participatory Designers: Zhan Yuanwang, Wei Zongquan

Project Area: 82 m^2

Major Materials: Stone, Rose gold, Wall covering and Wood finishes

本案承载着新时代的贵族精神，生活空间以低调、奢华的方式来表现。空间主要用棕色和金色来搭配，采用富有质感大理石装饰地面，运用了沉稳的色调作为背景色来演绎雅致的空间，空间的配色大气，符合业主的身份和品位。设计师还注重空间的细节设计，墙面上精心装饰的蝴蝶似乎在翩翩起舞，给空间带来一丝灵动的气息。

从空间整体的搭配到房间功能区的划分，空间的布局可谓是完美。设计师为空间选取了许多时尚、简约的家具来搭配，它们以独特魅力体现出了现代都市生活空间中闲适又兼具质感的特点。闪烁的水晶灯、大面积的装饰镜面、柔软的皮质沙发、细致的布艺抱枕等，设计师用低调而高贵的设计来突显空间的品质感，同时也彰显出了业主精致而有序的生活态度。软装的搭配是打造空间的重要手法，精致的家具配合棕色的大气稳重，再加上金色的优雅奢华，让整个居所充满了都市新贵的品位。

The captioned case bears the new ages' noble spirit. The life space is low-key and luxury. The main color of the space is brown and gold. The floor uses marble decoration. The background color uses steady color as a tone to show an elegant space. The space's color seems calm and matches the owner's identity and taste. Designers pay attention to the detail design of the space. The butterfly is dancing on the wall, bringing the space a sense of liveliness.

From the match of the space and the division of the functional area. The layout is perfect. Designers selects many fashionable and simple furniture to match with each other. They express unique charm of modern city life's leisure and beauty. Flashing crystal lamps, large decorative mirror, soft leather sofa, nice cloth pillow, etc., the designers use low-key and elegant space design to highlight the quality, meanwhile also highlights the fine and ordered attitude towards life. Soft decoration is an important technique to create the space; fine furnishings with brown stable atmosphere, coupled with luxurious elegant gold makes the whole city full of new noble taste!

摩登样板间 IV
欧式风尚

A 公寓

A Apartment

设计单位：金元门设计单位

设计师：葛晓彪

项目面积：179 m²

撰文：葛晓彪

摄影：刘鹰

Design Company: Gold Yuan Gate Design Company

Designer: Ge Xiaobiao

Project Area: 179 m²

Copy Writer: Ge Xiaobiao

Photography: Liu Ying

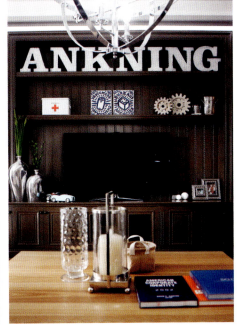

这是一户四口之家的房子，由于主人姓安，因此被设计师称为A公寓。设计师对原来的户型图进行了规划与改造，之前的户型有很大的问题，餐厅又小又窄，侧边的卫生间也很小，北阳台利用率低，还有一条很长的通道占据着客厅和餐厅、卫生间的面积。因此，设计师首先将朝西的卫生间和北阳台、餐厅进行了调整，这样一来使北面的卧房使用面积达到最大，空间也显得空阔。将设有落地景观窗的主卧卫生间调整

为书房，这样就能有效地利用自然采光的条件，并且，透过书房的通道使餐厅光线更加明亮了。设计师把中央空调主机放在西北朝向的卫生间旁，这样就能减少卧室的噪声。整个户型通过合理地改造后，餐厅的面积增大了。由于去掉了通道，房间的结构与功能也有所优化。

室内的实木部分采用了紫灰色与巧克力色相结合的色调，墙面采用浅灰色，地面选用的是自然的普通原木地板，给室内带来了温馨的气息。空间还结合了英国元素、文字图案、工业造型产品与金属色的物件，营造出一种具有现代都市简约、时尚特点的英伦风格。整体的设计让空间视觉达到较佳效果，不仅打造了艺术感和品位感，而且营造出了一种休闲、轻松的氛围。

主卧的顶棚设计考虑到中央空调的风口安装位置，采用了灯槽发光，整体的房间色调及装饰较为素雅，结合布艺的窗帘、吊挂的帘幔，使业主完全享受在非常清静的生活氛围中。男孩房的布置整体略显动感，而女儿房采用了较为传统的经典家具布置，这样也较符合业主的意愿。书房的设计较为简单，当暖暖的阳光洒满阅读空间，让人感受别有一番惬意。厨房的蓝色橱柜与白色蜜蜂图案瓷砖相互映衬，体现出英式传统的风格。

A公寓主要以设计与软装陈设为主，设计师贴心地规划布局并精心构筑每个细节，打造出了充满活力的空间。本案中的多数家具、摆设等都由设计师原创设计的，部分还亲自参与了制作，这样不仅节省了去各处采购的时间和精力，还为业主节省了不少费用，同时也取得了设计师想要的效果。

This is a houses for a four people family, since the master's name is Ann, so it is named A apartment by the designer. Designer re-plans and re-modifies the original size chart. The previous size chart has a big problem. The restaurant is small and narrow, the side bathroom is small, the north terrace utilization is low. And there is a very long passage occupying the living and dining rooms and toilet area. Thus, the designer adjusts the West toilet and North terrace and the restaurant, therefore, to maximize the use area of the north bedroom then the whole space seems wider. Besides, designer adjusts main bathroom into a study, so that we can effectively utilize natural lighting conditions, and, through the channel making the restaurant light more bright. The designer puts the central air conditioner on the northwest side of the bathroom, so that we can reduce the bedroom noise. Through rational transformation, the restaurant area is increased. By removing the channel, the structure and the function of the room has also been optimized.

The solid wood of the interior part uses purple gray and chocolate tones, the wall uses light gray, the floor uses natural common wood floor, bringing the room a warm atmosphere. The spaces use Britain elements, text pattern, industrial styling products and metallic objects, creating a modern metropolis with a simple, stylish features of the British style. The overall design allows the visual space to

achieve a better results, not only building the artistic sense and taste sensation, but also creating a casual, relaxed atmosphere.

Main bedroom ceiling design allows for the installation of central air conditioning vent position, using the light emitting groove, the overall tone of the room are simple and elegant, combining with fabric curtains, hanging drapes, so that the owner can fully enjoy the atmosphere of life quietly. The layout of boy room seems slightly sporty, the daughter room is decorated with traditional classic furnishings. This is also more in line with the wishes of the owner. Study design is simple, when the warm sun shines on reading space, people will have a pleasant experience. Blue kitchen cabinets and white bee pattern echoing with each other, reflecting the traditional British style.

The main part of the A apartment lies in the design and soft furnishings. Designer intimates layout and carefully constructed every detail, to create a vibrant space. Most of the furniture, furnishings of the case are originally designed by the designer, and some also personally involved in the production. It not only saves time and effort throughout the purchase, but also save a lot of costs for the owners, and gain ideal effect of the designer.

摩登样板间 IV
欧式风尚

陌上花开

Mountain Blossom

设计单位：大品装饰—DoLong 设计

施工单位：大品专业施工

项目面积：261 m²

主要材料：护墙板、石材、壁纸、地板、硬包

摄影：金啸文空间摄影

Design Company: Dapin Decoration — Dolong Design

Construction Company: Dapin Professional Construction

Project Area: 261 m²

Major Materials: Siding board, Stone, Wallpaper, Flooring, Hard pack

Photography: Jin Xiaowen Space Photography

MODERN SHOW FLAT IV EUROPEAN FASHION 065

本案以现代的设计理念勾画出欧式风格的空间效果,将简洁与复杂、质朴与华丽紧密融合、相互贯通。整个空间以白色为主色调,搭配栗色的家具、写实的配饰,打造出简洁而优雅的的生活空间。

根据业主的生活方式和居住人口的比例关系,设计师对原始户型进行了新的规划和定位,无论在空间利用上还是在房屋布局上,都对其按照舒适住所的要求进行了改造。在设计上,以浅色的壁纸搭配白色的木墙裙,米灰色的地砖搭配深色的走边线,使室内空间中软装与硬装相结合、深色与浅色相搭配,在视觉上具有对比感。在室内陈设上,设计师精心挑选了与业主身份、性格、爱好相吻合的家具及装饰品,古典风的吊扇灯与栗色的餐桌椅、沙发相呼应,恰到好处地营造出一个宁静、高贵、奢华的极致空间。

The captioned case uses modern design to outline European spatial style effects, tightly integrating and linking up the simple and the complex, the plain and the gorgeous. The entire space is in white color, matching with maroon furniture, realistic accessories to create a simple and elegant living space.

According to the owner's lifestyle and the ratio of the resident population, the designer re-plans and re-positions the original unit. No matter on the space utilization or the layout, the house has been transformed into more comfortable. The designer uses light color wallpaper to match the white wood paneling; beige gray tiles to match dark walk edges. Therefore, the soft and hard decoration in the interior space are combined together, with dark and light color forming a sense of contrast visually. On the interior furnishings, the designer carefully selects furniture and decorations that are suitable for the owner's identity, personality and hobbies. Classical ceiling fan and the maroon table and chair are echoing with each other, forming a peaceful, noble and luxury space.

摩登样板间 IV
欧式风尚

昆明中豪悦城花园样板房

Kunming Zhonghao Yue City Garden Show Flat

设计单位：云南渲染装饰工程有限公司

设计师：彭梅

项目面积：106 m²

主要材料：微晶石、全抛釉砖、天然石材、黑镜、油画镜框线、壁纸、定制墙画、马赛克

Design Company: Yunnan Xuanran Decoration Engineering Co., Ltd.

Designer: Peng Mei

Project Area: 106 m²

Major Materials: Ceramic stone, Whole cast glazed tiles, Natural stone, Black mirror, Framed oil painting line, Wallpaper, Custom wall paintings, Mosaics

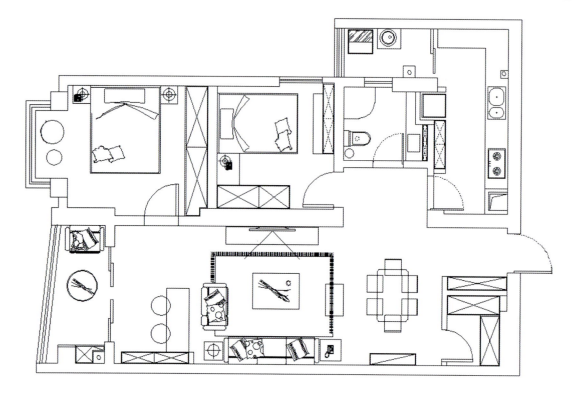

作为一个小户型空间，收纳及拓展空间是设计的核心。设计师将本案风格定位为后现代风格，因此在材质的使用上多选用了反光材质，这符合风格及空间的要求。在收纳上，本案着重体现了区域功能收纳多样化，有高大的鞋柜、大幅的卫生间镜柜设计，还有独立储藏室。阳台的收纳柜既遮盖了阳台上的管道，又利用了柱体及阳台管道之间的空隙，在满足阳台收纳的同时避免卫生死角，使业主在未来的使用中避免了无处收纳的尴尬境地。

在色彩的使用上，因本案业主为年轻女白领，而后现代风格材质及色彩又过于冰冷及坚硬，因此，空间中采用了柔软的皮料电视背景及过道纱幔，沙发背后使用浪漫紫红色彩及飘逸羽毛图案壁纸来点缀。鞋柜柜门的弧形线条及奥黛丽·赫本的黑白精致墙画与吧椅遥相呼应。个性化的沙发挂画，既弱化了空间过于冰凉坚硬的材质，同时也体现了业主作为都市女白领的时尚、干练及女性的柔美。在功能的布局上，吧台酒柜区弥补了原本客厅过长的缺点，同时业主在此处品着红酒听着音乐，浪漫而休闲。吧台的设计不仅丰富了业主的家居生活，也体现了其作为都市白领的高品质生活需求。

Being a small apartment space, storage space and development space is the core of the design. The designer design the case as a post-modern style building. Therefore, on the use of the materials he chooses reflective material which meets the requirements of style and space. As for the storage, the case emphasizes the regional functional multi-storage by tall shoe cabinet, large size mirror cabinet design, as well as separate storage room. Storage cabinets cover the pipeline on the balcony and take advantage of the gap between the cylinder and balcony pipeline. Design meets the balcony storage requirement while avoiding the health corner, so that avoid the embarrassment of storage in the future.

On the use of the color, since the owner of the captioned case is young female white-collar worker, and postmodern materials and colors are too cold and hard, therefore, designer adds soft leather on the TV background, waving curtain in the aisle and romantic purple color and elegant feather pattern wallpaper behind the sofa. The arc lines of the shoe cabinet door and Audrey Hepburn's black and white exquisite wall paintings echoes with the chair in the bar. Personalized sofa paintings weaken the cold and hard space, meanwhile reflects the owner, as an urban female white-collar, is fashion, capable and soft. On the layout, the bar area makes up the shortcomings of the too long living room. The owner here sips the red wine while listening to the music romantically and pleasant. The bar design not only enrich the owner's home life, but also reflect her high-quality requirements as an urban white-collar.

摩登样板间 IV
欧式风尚

有色生活

Colorful Life

设计单位：大品装饰—Dolong 设计
施工单位：大品专业施工
设计师：李启明
参与设计：周泽才
项目面积：220 m²

Design Company: Dapin Decoration — Dolong Design
Construction Company: Dapin Professional Construction
Designer: Li Qiming
Participatory Designer: Zhou Zecai
Project Area: 220 m²

刚开始接到这个案子的时候，其实业主已经找过设计同行连同父母的房子一起出过两套图纸了，他找我们公司是想签纯施工合同。经过和业主的沟通，才发现图纸中奢华的欧式风格并不是业主想要的。用男主人的话来说："我们需要的是'哇'的感觉。"个性十足的设计才是业主发自内心想要的。在不断地沟通中，我与业主达成一致，希望用一边玩一边设计的心态去表达空间，主题就一个字——"哇"！

前期的设计方案主要是对平面布局进行了修改，调整了部分空间功能，并通过参考大量的信息找到了业主喜欢的风格。女主人明显喜欢欧式结合美式的风格，还希望空间中添加女生都爱的闪闪发光的饰品。男主人却是喜欢用色大胆、充满个性的设计。他们都偏爱纯色系，注重各类具有创意的造型细节。我在设计中考虑了业主各自的爱好，决定用古典欧式的形态来塑造整体大框架，后期再加入纯色系乳胶漆，也充分利用了他们收集的各类小玩意来打造空间。

一直到最后快拍片前，依然在忙碌，做最后的软装设计。空间的效果我们都很满意，甚至超出了预期。空间就像是一个儿童乐园似的，这也是我们想要的模样，我们喜欢的丰富多彩的生活！

At the beginning of the case, the owner has found the designer who designs two sets of drawing for both the house of the owner and his parents. He wants to sign construction agreement with our company. After communicating with the owner, we finds he does not need the luxury European style. He needs a feeling of "Wa". He wants individual design style. During the communication, the owner and I reach an agreement. We decide to express the space full of fun. And the theme is one word——"Wa"!

The design theme at the beginning is modifying the layout and adjusting partial space function. Through a large amount of information we find the owner's favorite style. The hostess prefers the European-American Style. An also like to add blings to the space. The host likes colorful and individual design. They prefer pure color and pay attention to all kinds of creative styling detail. I think about the owner's interest and I decide using the classical European style to create the whole frame. Then I plan to add pure color latex paint and various small things to create the space.

We are busy to do the last soft decoration before the piece is done. Both of us are satisfied with the space effect, and even over the expectation. The space is like a children paradise. And this is also what we want as a colorful life!

摩登样板间 IV
欧式风尚

繁华之上·ECO城样板房

Over the Busy • ECO City Show Apartment

设计单位：福州 L&K 木可室内设计有限公司

设计师：李柯

参与设计：施家星、黄聪

项目面积：130 m²

摄影：周跃东

Design Company: Fuzhou L&K Mu Ke Interior Design Co., Ltd.

Designer: Li Ke

Participatory Designers: Shi Jiaxing, Huang Cong

Project Area: 130 m²

Photography: Zhou Yuedong

本案的样板房设计突显了作为展示空间的功能和装饰主义的美学，以时尚、新潮的设计理念，将随手拈来的新装饰元素融于其中，看似漫不经心的搭配，却深谙新古典的精髓意蕴。

设计师采用了乳白色、米黄色等浅色调，起到了在视觉上放大空间面积的作用。再搭配代表稳重、成熟的咖啡色和褐色，使空间整体色调和谐又富有层次，彰显出尊贵的空间格调。设计师除了通过色彩将焦点从餐厅转移到客厅之外，还选取了一些软装饰品丰富空间。餐厅在桌椅和餐具的选择上采用减法，用素雅的大色块使空间简单化；客厅却反其道而行，大胆采用加法，让毛毯、茶几、沙发层层叠加，利用形态各异的图案和格纹丰富家居场景。设计师注重家具与空间的风格统一，用优雅的曲线勾勒出家的品位。空间局部运用灯光和墙面造型把控室内气氛，精雕细琢的工艺与软装陈设显示了空间的高雅格调，体现出了业主的内涵。

The design of the show apartment highlights the aesthetic function as exhibition space and decoration space. The design ideal is fashionable with new decorative elements at random, which seems casual but contains the essence of new classical style.

Designers use milky white, beige and other light colors to visually enlarge the space area. And then they add stable, mature coffee and brown color, making the overall tone harmony and rich, and highlighting the noble space style. In addition to shifting the focus from the restaurant to the living room through color, the designers also select some soft furnishings to form an abundant space. The restaurant uses "subtraction" on the choice of furniture and tableware; uses elegant large blocks of color to make the space simplistic. While the living room is in the opposite direction by using "addition". The designers boldly pay attention to the superposition of blankets, coffee table and sofa, using different forms of pattern and check to rich the home scene. Designers will focus on the unified style of furniture and space with elegant curves sketch out the taste of home. Part of the space uses lighting and wall modeling to control indoor climate. Delicate technology and soft decoration display show the high-level taste of the space and the space's owner.

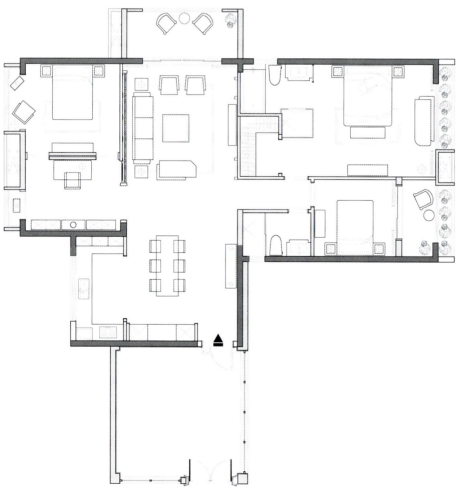

摩登样板间 IV
欧式风尚

武汉天地

Wuhan Space

设计单位：武汉郑一鸣室内建筑设计
设计师：郑一鸣
软装设计师：吴锦文
项目面积：218 m²
主要材料：大理石、装饰画、艺术壁纸、拼花地板、软包等

Design Company: Wu Han Zheng Yi Ming Interior Building Design
Designer: Zheng Yiming
Decoration Designer: Wu Jinwen
Project Area: 218 m²
Major Materials: Marble, Decorative paintings, Art wallpaper,
Parquet floors, Soft bag, etc.

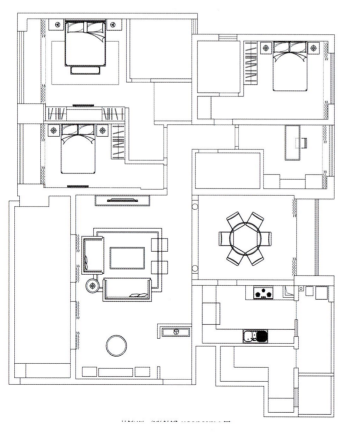

本案极力为空间营造一种温馨、阳光的欧式风情，硬装上强调华丽、雅致，没有复杂的镶金流银。但在家具配饰上，则把欧式风格的精致做到了极致。本案的用材强调环保，强调能够体现出家的温暖、阳光、舒适和休闲。遵循"少即是多"的设计理念，强调空间格调，彰显业主独特的品位。

步入室内，所处环境让人觉得一阵莫名的惊喜。深色边框搭配浅色布艺的家具，奢华中透着丝丝优雅和恬静。在注重装饰效果的同时，用现代的手法还原古典气质，让空间兼具古典与现代的双重审美效果，完美的结合也让人们在享受物质生活的同时得到了精神上的愉悦。

The captioned case strongly creates a warm, sunny European style. The hard decoration emphasizes gorgeous and elegant, with no complicated gold or silver. However, on the furniture accessories, designer uses extreme European-style. The case emphasizes environmental protection, stressing the warmth, sunshine, comfort and leisure of the family. Following the "less is more" design philosophy, emphasizing space style, highlighting the owner's taste.

Entering into the room, the environment makes people feel a burst of inexplicable surprise. Dark border matches with light fabric furniture seems luxury, elegant and quiet. Using modern techniques to restore classical temperament while emphasizing decoration effect, so that the space has both classical and modern dual aesthetic effect. The perfect combination makes people enjoy material and spiritual life.

Central Park House

设计单位：台北玄武设计

设计师：詹皓婷、萧洛琴

软装设计师：胡春惠、杨惠涵

项目面积：92 m²

摄影：赵志成

撰文：程歆淳

Design Company: Taipei XuanWu Design

Designer: Zhan Haoting, Xiao Luoqin

Decoration Designer: Hu Chunhui, Yang Huihan

Project Area: 92 m²

Photography: Zhao Zhicheng

Copy Writer: Cheng Xinchun

MODERN SHOW FLAT IV EUROPEAN FASHION 099

纯色背景 鲜明点缀

本案呈现了玄武设计在物质与心灵、个人与家庭、东西文化之间求取的微妙平衡。步入玄关，中式结合装饰主义的金属墙面熠熠生辉，远观可见平滑质感，近看、细节之精致让人震慑，两只西洋棋摆饰分置左右，彷佛透露几许童心。以黑、白铺底的客厅空间里，设计师只拣取少量明色（鲜绿、亮银）点缀其间，并利用上繁下简线条收束视觉，一如业主低调的性格。也可利用绒毛地毯、缎布抱枕、窗帘等堆栈出温暖感，颇有安居乐业的韵味。餐厅背景改采棕色木皮，大幅画作反照出餐厅悬吊的水晶灯，和餐椅下摆流苏、客厅水晶隔帘一起，形成一气呵成的奢华感。

独善共乐 空间嬉游

从餐厅步入阅读空间，体现着家庭成员对于个人生活的向往。设计者选择简练线条桌椅，石材桌面、搭配一只跃动感雕饰，于动静之间，酝酿无限巧思。步入长廊，分置左右的主卧和其他卧室，均以素白为底，配合深浅不一的纯黑、浅灰材料，彷佛缓慢推进的背景音，金棕灯饰、镶银抱枕，与猛然跃出的金黄窗帘、草绿地毯和桃红被褥，鲜明地诠释了业主的个性。大量收纳空间被巧妙隐藏于转角、壁面之后，或是一体二用——让衣帽间同时具备储物功能。

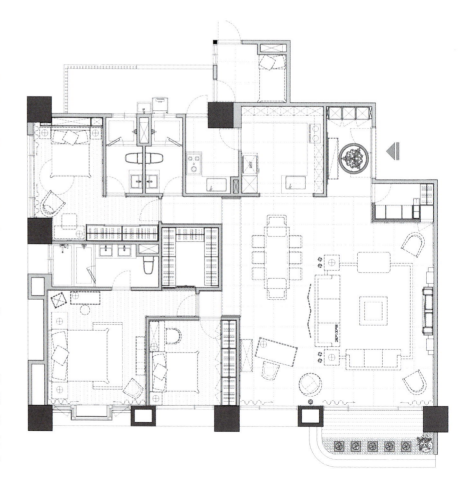

Pure Background Bright Ornament

The case presents the delicate balance between material and heart, individual and family, Eastern and Western cultures of XuanWu Design. Stepping into the entrance, Chinese and decorative style metal wall is shiny, looking from the distance one can see the smooth texture, while looking closely, the details are astonishing. Two chess decorations are at left and right seems like the childhood's heart. In the black and white living space, designer just picks a few bright colors (bright green, silver), then uses complicated up line and simple down line to limit the vision, as the owner's low-key personality. Besides, floss carpets, satin pillows, curtains and other materials can create warm feeling and charm lively sense. Restaurant background uses brown wood cover; the large paintings reflect restaurant's suspended crystal lamps and chairs' edge hem as well as living room's crystal curtain, forming coherent luxury sense.

Playing Happily In The Space

Walking from the dining room into the reading room, the families are yearning for personal life. Designer uses concise lines of furniture, stone table, with a jump feeling decoration, brewing unlimited ingenuity between the movement and quietness. Walking into the promenade, the master bedroom and other bedrooms use pure white at the bottom. The dark and light black color and light gray color materials are like slowly songs going on. Brown lights, silver pillows and jumping golden curtain, green carpet and pink bedding shows the owner's individuality. A lot of storage space is cleverly hidden in the corner, after the wall, or being one-body-two-use——having the storage capabilities at the same time.

摩登样板间 IV
欧式风尚

静和之美

The Beauty of Peace

设计单位：大品装饰—DoLong 设计
施工单位：大品专业施工
项目面积：142 m²
主要材料：文化石、复古砖、壁纸
摄影：金啸文空间摄影

Design Company: Dapin Decoration — Dolong Design
Construction Company: Dapin Professional Construction
Project Area: 142 m²
Major Materials: Culture stone, Tile, Wallpaper
Photography: Jin Xiaowen Space Photography

静，是一种大美，一种深沉的美，一种更富感染力的美，一种恒久的美！心静自然凉，表述的是沉静之美。你只有心灵沉静，不浮躁、沉住气，才会心态平衡、心平气和。在这个喧嚣的世界，我们需要能够独享清静的一隅，笑看世间百态。

每个人对家的需求各不相同。"温和从容，岁月静好"，虽然寥寥几字，但已经反映了业主对于自己家的定义。温馨、安静、舒适，看似简单的需求，却对设计师有着更高的要求，通过反复沟通，确定了设计理念及家具样式，一个具有美感的空间就此应运而生。

由于楼层的原因，室内的采光相对较弱，因此客厅大面积采用了明亮的色调，其他空间在局部以跳跃的蓝色、明黄色丰富空间。书房和小孩房则运用了橄榄绿色和蓝色，颜色的区分使得每个空间都有其独特的属性。考虑到女主人喜欢花草，在设计及软装搭配过程中，将花草以不同的形式存在于整个家中，为空间带来些许自然气息。植物的点缀不仅体现了业主怡然自得的性格，也使得家成为了一个轻松、自在的地方。

Peace is a great beauty, a deep beauty, a more movable beauty, a long-lasting beauty! Peaceful mind brings cool feeling which expresses the beauty of quiet. People with peaceful mind are quiet, not impetuous and calm. In this noisy world, we need an exclusive quiet corner, laughingly facing all the phenomena in the world.

Everyone has different need of home. "To Be Gentle and Calm, Quiet Days is Good." The few little words reflects the definition of the owners to their home. Warm, quiet, comfortable seems simple but has higher requirements to the designer. Through repeated communication, they decides the design ideal and furniture style, then an aesthetic space has emerged.

Because of the floors, interior lighting is relatively weak, so the large living room area uses the bright colors. The other space uses jump blue, bright yellow to make the space abundant. Study and the kids room uses olive green and blue, the color distinction allows each space has its own unique properties. Considering the hostess like flowers, in the match of design and soft decoration, designer put flowers in different forms throughout the home, bringing the space a little natural flavor. Ornament plant not only expresses the owner's peaceful character, but also makes the house a relaxing, comfortable place.

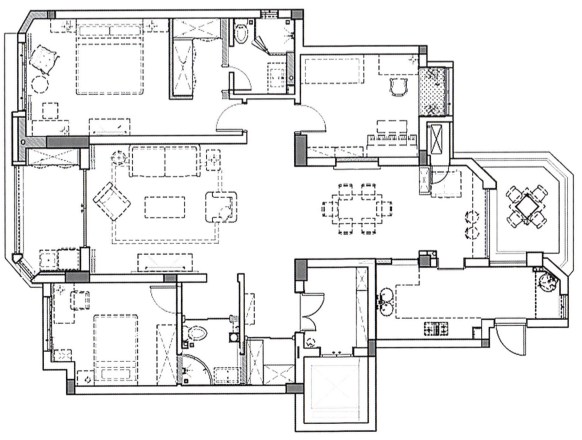

摩登样板间 IV
欧式风尚

世茂臻园

Shi Mao Zhen Yuan

设计单位：筑易设计

设计师：郭娟

项目面积：198 m²

主要材料：木纹石、壁纸、仿古砖、复合大理石拼花、银镜

摄影：施凯

Design Company: Zhu Yi Design

Designer: Guo Juan

Project Area: 198 m²

Major Materials: Wood stone, Wallpaper, Antique tiles, Composite marble mosaic, Silver mirror

Photography: Shi Kai

MODERN SHOW FLAT IV **EUROPEAN FASHION** 111

香槟玫瑰

客厅空间中以大地色系为主色调，适当地添加了一抹鲜艳色彩，走进这里，能感受到像奶油色般融化的甜腻芬芳，让人沉醉。室内空间由大面积的暖色调作为主导，搭配巧克力色的沙发垫、地毯作为辅助，亮白色与奶油色的融合，使空间色彩自然衔接，起到过渡的作用。

设计师在空间中采用金色的奢华装饰品点缀，烘托出家居的优雅与和谐，将业主想要拥有的豪华与浪漫尽情展现。在前期设计中，业主与设计师沟通良久，在设计的过程中，设计师特意增添了业主对美好独特理解与感悟的细节，最终呈现出符合业主身份与品位的、独一无二的高雅居所。

Champagne Rose

Living room space's main color is earth color, the minor color is bright color. Coming into the space, one can feel a sweet cream-melt fragrance, making people intoxicating. Interior space is consist of a large area of warm color with chocolate sofa cushion and carpet. Fusion of bright white and cream color plays a role of transition, making the space's color connected naturally.

Designer uses golden luxury ornaments in the space, expressing an elegant and harmonious home furniture atmosphere, displaying the owner's luxury and romantic wish. In the preliminary design, the owner and designer communicate for a long time. Designer deliberately adds the owners' good and unique understanding of the details and insights, and finally presenting an unique and elegant accommodations.

摩登样板间 IV
欧式风尚

海航城E户型样板房

Hai Hang City E Unit Show Apartment

设计单位：深圳太合南方设计事务所

设计师：王五平

项目面积：165 m²

主要材料：银镜、木地板、壁纸、全抛釉砖

Design Company: Shenzhen Taihe South Design Firm

Designer: Wang Wuping

Project Area: 165 m²

Major Materials: Silver mirror, Wood flooring, Wallpaper, Polished glaze tile

本案展现出现代、精致、简约的空间气质，与业主所要求的一致。温馨明亮的空间、高贵典雅的白色木线、时尚奢华的家居，墙面上纯色与隐花相融相成，无不透出业主对生活完美极致的追求，并在品位中渗透着对生活的热爱。在平面布局上，设计师通过对墙体的改动，实现了空间的共享，使业主的行走路线更加流畅，提升了空间利用率。在墙面的设计处理上，设计师通过使用大面积的镜面，以达到放大空间的视觉效果。

女主人对轻奢风格情有独钟，所以设计师灵活使用了薰衣草色，这种神秘又优雅的色调在家居饰品中得到淋漓精致的展现。高贵与雅致、时尚与魅惑，都是设计师想表达的特质。除了在家居饰品中所体现的奢华之外，业主还有着一种对自然气息的向往和追求，琴房里的那幅白桦林的挂画足以见证。

The captioned case shows a modern, sophisticated, simple space temperament, as the owner's requested. Warm and bright spaces, elegant and noble white wood line, luxury and fashionable household, and the solid and hidden flowers on the wall reveal the owner's pursuit of perfect life style and the love towards life. On the layout, the designer achieves the share of the space by changing the wall so that the route is smoother, the space utilization is improved. On the design of the wall, the designer visually enlarges the space through the use of mirror.

Mistress loves light extravagant style, so the designer flexibly uses lavender, this mysterious and elegant color has been used on various exquisite jewelry in the home. Noble and elegant, fashionable and charm are all what the designer wants to express. In addition to the luxury addition reflected in the home accessories, the owner also pursuits a kind of natural flavor, and the birch paintings in the piano room acts as a witness.

摩登样板间 IV
欧式风尚

丰泽园一期

Feng Ze Yuan First Phase

设计单位：金元门设计单位

设计师：葛晓彪

项目面积：172 m²

主要材料：大理石、墙布、实木板材、瓷砖

撰文：葛晓彪

摄影：刘鹰

Design Company: Jin Yuan Men Design Company

Designer: Ge Xiaobiao

Project Area: 172 m²

Major Materials: Marble, Wall covering, Solid wood, Ceramic tile

Copy Writer: Ge Xiaobiao

Photography: Liu Ying

蓝调的优雅

业主是位有小资情怀的女孩，一直梦想着有一套梦幻典雅而又简约不凡的理想居所。设计师通过沟通并结合自己的独特创意，为其量身定做了这套都市新贵般居室空间，将原本空洞乏味的空间变得丰盈充实，以一种鲜活的姿态呈现在她的眼前。

设计师首先考虑到的是公共区域的整体配置，色彩上主张明快又简洁的配色，让整体家居环境充满了浪漫、温情的气息，使人一进公寓便感觉如沐春风、沁人心脾。玄关和阳台处的玻璃隔断设计，透过自然光影的折射让整个房子通透、明朗。阳光洒入屋内，经过墙壁和地面散射在整个空间，使空间隐约透露着朦胧感。客厅彰显出简约大气，白色系的护墙、深色系的沙发相互搭配，墙面、装饰画相得益彰，赋予了空间平衡之美。而玻璃茶几上具有艺术气息的芭蕾舞女孩，玻璃储物罐等零星的点缀，恰到好处地显示出业主崇尚的

简洁与优雅,摒弃了过多的烦琐和奢华。餐厅用少许的蒲公英配饰作为点缀,和紫色的餐椅相互烘托,使整个空间看起来温馨、简约。卧室的设计以明亮、恬静为主,对色彩的运用给人温和、柔美的感受。主卧白色的护墙配上深紫色系的壁纸、窗幔,温馨而又浪漫,用浅色的布艺作为点缀,越发显得悠闲、雅致。客卧的主题是清新,盘子作为装饰画出现在主题墙,配合蓝白格子壁纸,富有艺术气息。整体空间的设计充满了自在与情调,构成了具有文化意味的生活方式,让业主能享受美好的蓝调生活,体会属于自己的静谧时光。

Blues' Elegance

Owner is a girl with petty feelings who is dreaming to have a set of extraordinary and elegant but simple ideal accommodation. Through communication, designer uses his unique creativity to build a set of new urban style living space, making the original empty space becomes an abundant and rich house, freshly showing in her eyes.

First of all, designer considers the overall style of the public areas. The color uses bright and simple color, so that the overall home environment is full of romantic and warm atmosphere, people feel in a spring wind when stepping into the house. The glass partition design at the entrance and the balcony make the whole room transparent and clear through the reflection of the natural light. The sun shines into the house, through walls and floors scattering throughout the space so the space vaguely reveals a hazy sense. The living room highlights simple and wide atmosphere, the white wall, dark sofa match with each other, the wall and the decorative paintings complement each other, giving the space a balance beauty. The sporadic ballet girl stands on the glass coffee table, together with the glass storage show the simplicity and elegance of the owner, abandoning too much cumber and luxury waste. The

owner, abandoning too much cumber and luxury waste. The restaurant uses a little dandelion accessories as ornament, setting off with the purple chairs, so that the whole space looks warm and simple. Bedroom design bases on bright and quiet, the use of color gives people a sense of gentle, soft feeling. The white wall of the main bedroom is coupled with dark purple wallpaper and curtain seems warm and romantic. The light-colored cloth seems relaxed and elegant. The theme of the guest room is fresh, the dishes are now shown as a decorative theme of the wall, the blue and white squares wallpaper is full of art.

The overall design is full of comfort and ambiance, constituting a new life style with cultural meaning. Therefore the owner can enjoy the fine blues style life, experiencing her own quiet time.

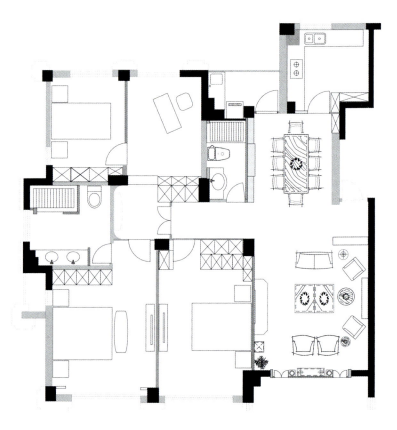

摩登样板间 IV
欧式风尚

深圳海航城 C2 户型样板房

Show Flat for C2 House Type, Shenzhen HNA City

设计单位：深圳太合南方设计事务所

设计师：王五平

项目面积：85 m²

项目地点：广东深圳

主要材料：木地板、壁纸、玫瑰镜钢、木纹灰全抛釉砖

Design Company: Shenzhen Taihe Nanfang Design Firm

Designer: Wang Wuping

Project Area: 85 m²

Project Location: Shenzhen Guangdong

Major Materials: Wood Floor, Wallpaper, Rose Mirror Steel, Wood Grain Gray Glazed Ceramic Tile

现代、精致、时尚和奢华，是本案要彰显的风格；空间的宽敞感和清晰的层次感是本案要达到的效果。在空间改造上，通过对墙体的改动，设计团队扩大了空间的面积，使空间在视觉上显得开阔而又敞亮。在餐厅立面的处理上，通过镜面来点缀，空间层次更加清晰分明，并有效地拓宽了空间。

在材质搭配上，设计师追求精益求精，运用了黑色拉丝不锈钢踢脚线、玫瑰金钢线条和皮质硬包等，使空间显得高雅而又奢华。在家具的搭配上，以款式新颖、造型独特、现代时尚为主导，彰显主人高品位、高质量的生活。

This project tries to demonstrate the style of being modern, exquisite, fashionable and luxurious. This project is required to display the effects of expansive space and clear layers. As for space regeneration, through changes towards the wall, the design team expands the area of the space, making the space appear broad and bright visually. As for treatment towards the facade of dining hall, through embellishment of mirror surface, the space layers are made clearer, efficiently expanding the space.

As for collocations of materials, the designer tries to make everything perfect, with black stainless steel foot margin, rose gold steel line and leather hard rolls the space is made elegant and luxurious. As for furniture matching, the designer focuses on novel model, peculiar format and trendy appearance, displaying high taste and high quality life of the property owner.

摩登样板间 IV
欧式风尚

雅奢素影

Ya She Su Ying

设计单位：孔•设计工作室

项目面积：190 ㎡

项目地点：湖北武汉

主要材料：环保水性漆、玻化砖、石膏线、新古典家具等

Design Company: KONG•Design Studio

Project Area: 190 m²

Project Location: Wuhan Hubei

Major Materials: Environmentally friendly water-based paint, Tiles, Plaster, Neo-classical furniture, etc.

"雅"代表雅致,"奢"代表奢华,"素"代表素净,"影"代表生活,这是本案设计的精髓。案例中以银色勾勒线条,以白色为背景色,干净、素雅,表现出业主对浪漫生活的甜蜜憧憬。

从功能性讲,此户型四房两厅两卫带大阳台,由于阳台面积很大,所以被一分为二做成榻榻米阳光书房和生活阳台,用帘子作为隔断,并不影响空间的采光通风。客、餐厅中间的过道较大,正好满足业主摆放三角钢琴的需求。业主为四口之家,夫妻留学归来,有两个小孩,有些西方的生活习惯。女主人会常在家给小孩做烘培,由于厨房只能满足中餐厨的需求,所以在餐厅区加入了一排西厨柜。两个小孩目前都在上幼儿园,暂时共住一间房。练功房因为做了榻榻米、书柜、书桌,也可以当书房用。主卧满足了业主储藏的要求,不仅有一排满墙衣柜,连主卫也改成了衣帽间。

室内空间的主色调为紫色,无论是淡紫色的墙壁、新古典家具,还是柔和的织物装饰,都营造出安静、浪漫的氛围。

"Ya" represents elegant, "She in Chinese" represents luxury, "Su" for simple, "Ying"for shadowy, which is the essence of the design. Case uses silver outlining lines, white as background color, clean, simple and elegant, showing the owner's romantic life sweet dream.

From the function,the unit has four bedrooms, two living rooms and two bathrooms.Since the balcony has large area, it is divided into two parts, one for tatami sun study, the other is living balcony. The curtain acts as a partition which does not affect lighting or ventilation. Passenger between the living room and dinning room are large enough to put a grand piano. The house is designed for a four member family——a couple returned from abroad with two children. They have some western habits. The hostess often bake at home for the children.Because the kitchen can only meet the needs of Chinese food, therefore adding a row of west cabinets to dinning area. Two children are currently in kindergarten, temporarily sharing a room. Training roomhas tatami, bookcase, desk, may also be used as study. Master bedroom meets the storage requirements of the owner;ithas a row of wall closet, even the main bathroomhas changed into cloakroom.

The interior's main color is purple, either lavender walls, neo-classical furniture, or softdecorative fabric, all creating a peaceful, romantic atmosphere.

摩登样板间 IV
欧式风尚

蓝羽雀

Blue Finch

设计单位：福州瑞亚装饰设计工程有限公司

设计师：林诚

主要材料：瓷砖、大理石、水曲柳原木、壁纸、有色涂料、木饰面板、玻璃、实木地板

摄影：黄访纹

Design Company: Fuzhou Ruiya Decoration Design Engineering Co., Ltd.

Designer: Lin Cheng

Major Materials: Tile, Marble, Ash logs, Wallpaper, Colored paint, Wood panels, Glass, Wood flooring

Photography: Huang Fangwen

时尚是一时的，优雅却是永恒的。它固然与财富、地位有关，但不是单纯依靠奢华就可以包装得出来，它离不开岁月的沉淀和生活的历练，更离不开一定的文化艺术修养。

本案为了体现生活的本质和简单的韵味，特地选用质朴的元素和材质进行演绎。为达到更高层次的家居艺术表现，设计师选择了"空灵"和"静美"为核心理念来营造这个家的独特意境。它的配色和造型简约而纯净，有种超凡脱俗的典雅气质。它洗尽了生命的铅华，远离了喧嚣，独守住内心的安宁。它摒弃华丽的装饰，以直线、纯色来缔造空间的气韵，使空间回归本真，回归简单，把斑斓的万物凝练成眼前隽永悠长的景象，通过别具匠心的搭配和组合，让单纯、优雅、自然且充满个性的生活氛围跃然眼前。

Fashion is temporary, but elegance is timeless. It is of course related to wealth, status, but it can not simply rely on the luxury package. It cannot be separated from time's sedimentation and experience in life and also a certain cultural and artistic accomplishment.

In order to reflect the nature and the simple charm of life, the designer specially chooses rustic elements and materials for interpretation. In order to achieve a higher level of artistic expression at home, the designer selects "ethereal" and "quiet beauty" as the core concept to create a unique mood of the family. Its color and shape is simple and pure, with a kind of unearthly elegance. It washes away the magnificence of life, being away from the hustle and bustle, and holds the inner peace. It abandons ornate decoration, creating the charm of the space in a straight line, a solid color, making room return to the truth and simple. The gorgeous things have been condensed into a meaningful long scene, through the ingenuity of combinations, the simple, elegant, natural atmosphere of life vividly appears.

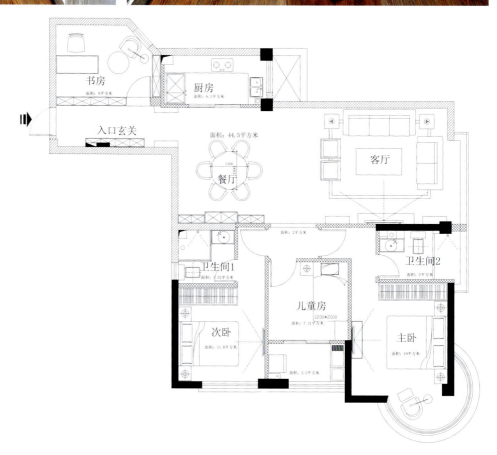

摩登样板间 IV
欧式风尚

白色恋歌

White Love Song

设计师：林锦

主要材料：木饰面、壁纸、金刚板、镜面

摄影：李玲玉

Designer: Lin Jin

Major Materials: Wood finishes, Wallpaper, Diamond plate, Mirror

Photography: Li Lingyu

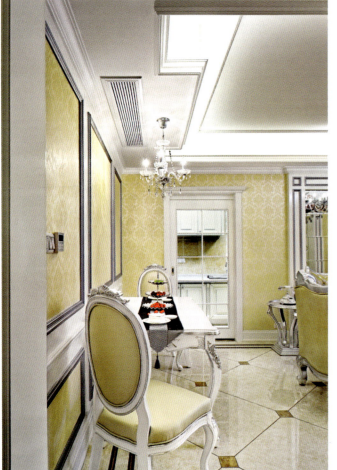

浅色调的简欧风格一般比较受年轻人的喜爱，本案的业主就是一个年轻、时尚的美女，自身比较喜欢传统的简欧风格，所以设计师和她沟通了之后就定位成这样一种色调，既有简欧的奢华，又不失青春、时尚的气息。

本案试图营造一种天真烂漫又时尚的感觉，因此选择了米白色作为主色调，搭配以银色的线条，再加上温馨的米色调壁纸。整个空间没有过多的色彩，整体色调柔和而细腻。在造型上，设计师也没有用过多的造型元素来阐释其理念，而是用类似的设计符号贯通不同的空间，在统一的风格中加了些许细腻的变化。为了保持装饰元素的一致，橱柜和衣柜也是特别定制的。主卧紫色的浪漫情怀，次卧香槟色的柔和情调，共同为室内谱写了一首清新、柔美的歌谣。

Light colored simple European style is generally loved by young people. In the captioned case, the owner is a young, stylish beauty who prefer traditional simple European style. So that after communication, the designer positions the space in such a tone which is luxury, young and stylish.

The captioned case try to create a sense of naivete and fashion, so designer chooses beige as the main color, then add silver lines, warm beige wallpaper. The whole space has no too much color, the overall tone is soft and delicate. In shape, the designer does not use too many design elements, but with similar design symbols through different spaces, adding delicate changes into unified style. In order to keep the decorative elements same, cabinets and wardrobes are also special made. Purple romantic of the main bedroom, champagne and soft mood of the second bedroom composing a piece of fresh and soft song.

醇美乐章

无需花哨的色彩，浅浅的巧克力色营造出温馨的家的氛围，厚实的归属感在家中萦绕，似乎有浓郁的甜香味在空间中弥散。设计师的构思是去掉堆砌的颜色和摆设，去掉繁复设计的多余家具，用简单的线条来打造大气的空间。统一色调的墙面作为客厅的主打，带给人柔和的感受，无论是壁纸还是织物或是地板和柜子，相互交织使整体感觉非常的活跃让人回味。

居住空间作为业主展现自我的背景舞台，必须充分展现其生活习惯与精神追求的特质。而这也是贯彻每一处空间场景的设计主轴。空间中随处可见的轻松的纹样、精致的饰品、柔美的布艺，将空间合奏出最幸福的乐章。而香槟金的点缀让空间具有感性的柔和、低调贵气而不张扬，开阔舒适又不失温情，是对品质生活的优雅态度，使业主能收获满满的幸福和感动！

Wonderful Music

Without gaudy color, the light chocolate color creates a warm family atmosphere; thick sense of belonging lingers at home; strong sweet scent disperses in the space. The idea of the designer is to remove the excess complicated colors and furnishings, using simple lines to create wider space. Unified-color wall is the tone of the main living room, bringing a soft feel to people. No matter wallpaper or fabric or flooring or cabinets, the intertwined active feeling is very impressive.

Living space, as the owner's background stage must fully demonstrates his living habits and characteristics of spiritual pursuit. This is also the main axis of every design. Relaxed patterns, fine jewelry and soft cloth of the space compose the happiest chapter. The gold champagne ornament emotionally softens the space which seems noble but not publicized, open, comfortable and warm. It is also an elegant attitude towards life, so that the owner can harvests full happiness and emotion!

摩登样板间 IV
欧式风尚

宅心物语

Homey Heart Story

设计单位：昶卓设计黄莉工作室
施工单位：怡明施工
软饰提供：昶卓软饰
项目面积：262 m²

Design Company: Chang Yong Design Huang Li Firm
Construction Company: Yi Ming Construction
Soft Decoration Provider: Chang Yong Soft Decoration
Project Area: 262 m²

进门第一眼，映入眼帘的是大气的走道，走道左边是餐厅，右边是客厅，功能分区非常完整。走近一看，你会发现这个空间的搭配非常协调，这是因为沙发的皮料选择、电视背景墙的硬包以及窗帘都是经过设计师精心考量的结果。在餐厅的设计中，将原来的飘窗拆掉以后，做了弧形的地柜，与餐桌、顶面、地面浑然一体。厨房经过合理改造后，现在是中西合并的综合厨房，厨房中的小小的岛台现在成了最方便实用的地方。餐厅中，飘窗左右的高柜全部都能用来收纳，这个隐藏的的储藏空间不太显眼却很实用。蓝色系的男孩房布局合理、色调温馨，让孩子在知识的海洋里遨游。从卧室空间看向客厅，你会发现处处都是用心的设计。

Stepping into the door, the first glance is the wide and bright aisle, on the aisle's left is the restaurant, on the right is the living room, the functional area is very comprehensive. Taking a close look, you will find the space very well coordinated. Because the selection of leather of sofa, the TV background hard pack and curtains are carefully considered by the designer. In design of the restaurant, the designer removes the original windows, and makes a curved floor cabinet which become a integration together with the dining table, the top surface and the ground. Kitchen, after reasonable transformation, now is a combination of Chinese and Western style. The small kitchen island bench becomes the most convenient and practical place. In the restaurant, high windows and cabinets can be used for storage. This hidden storage space is inconspicuous but very practical. Blue son room layout is reasonable, the color is warm, so that children can travel in the ocean of knowledge. Looking from the bedroom to the living room, you'll find every corner is carefully designed.

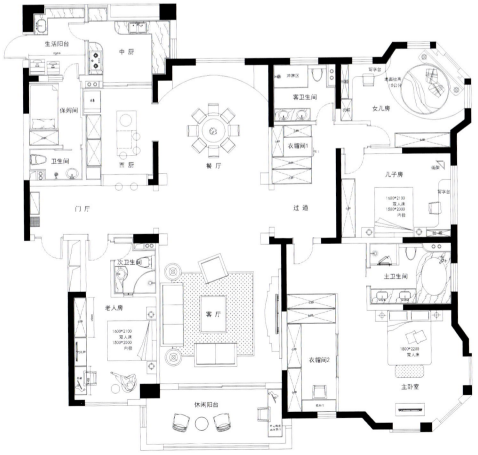

摩登样板间 IV
欧式风尚

尖沙咀名铸

Tsim Sha Tsui Ming Zhu

设计单位：Danny Chiu Interior Designs Ltd.

设计师：赵智铭（Danny Chiu）

项目面积：202 m²

项目地点：香港

主要材料：木纹大理石、马赛克、清镜

Design Company: Danny Chiu Interior Designs Ltd.

Designer: Danny Chiu

Project Area: 202 m²

Project Location: Hong Kong

Major Materials: Wood marble, Mosaic, Clear lens

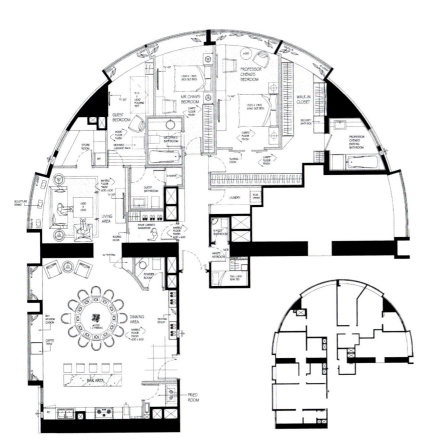

MODERN SHOW FLAT IV **EUROPEAN FASHION** 187

这个房子由两个空间打通，偌大空间划分成两大区，左边是满足宴客需要的宴会厅及客厅，右边是业主的卧室，两者分隔明确。空间设计以女主人的喜好为先，采用了深浅不一的柔美紫色，衬以清新白色，营造出淡雅、浪漫的氛围。古典的家具、随处可见的花纹图案，给人一种身处欧洲贵族大宅的感觉，高贵却不俗气。

应业主的要求，宴会厅中央放置了一张8～12人使用的餐桌，设计师以极富线条美的弧形电视墙调节了宴会厅的方正格局，使空间更流畅、平顺。还采用了镜钢框围绕灯槽和水晶吊灯，呼应地面的灰木纹云石圈，勾勒出视觉焦点。另一焦点是冷气槽的设计，沉闷的出风口变成典雅的花纹图案，还有淡化横梁的效果。与之呼应的是电视墙及吧台下白紫相间的马赛克玫瑰花纹，设计精致细腻。

电视墙以珠光白色、紫色马赛克拼砌玫瑰花图案，将客厅门隐藏其中，空间统一又大气。客厅则以紫色元素搭配。深紫色绒布沙发造型古典，雕花精致细腻，与紫色水晶灯一起散发着高贵气息。电视墙以方格线条为主，两旁衬以清镜，高贵中透露出简约、时尚，现代融合着古典，突显出层次丰富。

The house has been got through by two spaces. The huge space is divided into two areas; the left is the living room to meet the ballroom and banquet needs; the right is the owner's bedroom. The two parts are separating clearly. Space's design follows the hostess's first wish, using light and deep soft purple and lining with fresh white to create an elegant and romantic atmosphere. Antique furniture and patterns give people a sense of being in the European nobility mansion which seems noble but not tacky.

According to the owner's request, the central ballroom has been set a dining table for eight to ten people. Designer uses curved beautiful arc TV wall to regulate the ballroom structure, making the space more fluid and smooth. The designer also uses a steel frame around the lamp trough and crystal chandeliers to echoes the ash wood circle marble floor, and sketches out a visual focus. Another focus is the design of cold air ducts. The boring outlet changes into elegant patterns with desalination beam effect. And the match is the delicate designed white and purple color mosaic pattern under the television and the bar counter.

TV wall uses pearl white and purple mosaic to make rose pattern, hiding the living room door in it, making the space unified and generous. Living room is decorated with purple color. Purple velvet sofa is classical style with carved, delicate pattern, exuding nobility together with the purple crystal lamp. TV wall is grid lines oriented, both sides are lined with clear lens, exposing noble, simple, stylish and modern classical fusion, highlighting the rich layers.

Baoli Saina Vera X Show Apartment

设计单位：广州市韦格斯杨设计有限公司

项目面积：93 m²

项目地点：广东广州

主要材料：西班牙米黄大理石、爵士白大理石、黑白根大理石、灰色镜面、软包、羊毛地毯

Design Company: GrandGhostCanyon Designers Associates Ltd.

Project Area: 93 m²

Project Location: Guangzhou Guangdong

Major Materials: Spain beige marble, Jazz white marble, Black and white roots marble, Gray mirror, Soft bag, Wool carpets

本案的设计理念来自英伦风格的延伸，设计师以现代简约的手法勾勒出英伦的氛围。本案户型面积为93 ㎡，是三房二厅二卫二阳台的户型结构，平面布局方正，空间动线流畅，动静分隔适宜，户型非常实用。

设计师运用了西班牙米黄大理石、爵士白大理石、黑白根大理石等主要石材，选用灰色镜面、软包、羊毛地毯等材料，以欧式简约的表现手法来演绎空间。无论从细节上展现了欧式的艺术氛围，营造出一份浪漫而温馨的优雅气质。

空间一系列的米色调配搭、简单的英伦风格、Burberry 的格子元素，都在软装的细节上有所体现。英式的街头文化、英式著名的建筑物及历史文化等挂画及挂饰，映衬出浓郁的英国色彩。设计诠释了时尚、尊贵的高品质生活，丝丝入扣的细节表达，丰富的空间层次，都表现出设计师为业主提供高端、完善生活的衷心诚意。

The design concept of the captioned case is the extension of British style. Designer uses modern simple approach to sketch out the British atmosphere. The house area is 93 m^2 which contains three bedrooms, two living rooms, two bathrooms and two balconies. The layout is square shape with smooth generatrix. The space has reasonable and practical structure.

Designer uses Spain beige marble, jazz white marble, black and white root marble and other major stone, then uses gray mirror, soft bag, wool carpets and other materials to interpret European simple space, detailed showing the European arts scene and creating a warm and romantic elegance atmosphere.

The beige colors in the space, the simple British style, and Burberry plaid elements are reflected in the details of the soft decoration installation. British street culture, British famous buildings and historical culture, paintings and ornaments set off deep British colors. The design interprets fashionable and noble high quality of life. Detailed expression and rich space level show the heart of the designer to provide a high-end and better life.

Fuzhou Rongxin Dawei City Villa

设计师：蔡奇君、洪燕辉

项目面积：450 m²

主要材料：石材、瓷砖、实木、智能家居

摄影：黄访纹

Designers: Cai Qijun, Hong Yanhui

Project Area: 450 m²

Major Materials: Stone, Tile, Whole wood, Smart home

Photography: Huang Fangwen

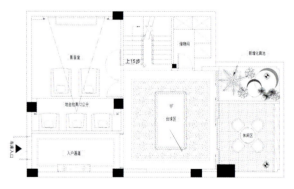
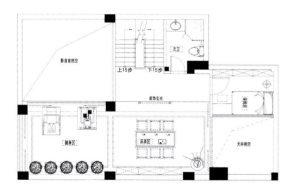
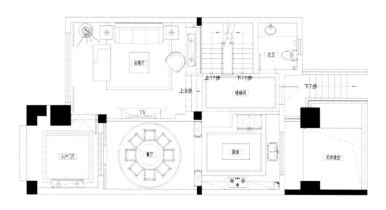
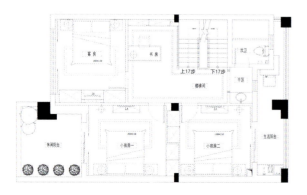

本案为简欧风格，客厅采用大面积米色大理石装饰，色调明亮，给空间带来一丝温馨。在家具造型方面，欧式线条的柔美，让整个空间看上去轻盈浪漫、富有诗意。餐厅采用了沉稳的色调，显示出奢华的一面，再搭配高调的软装饰品，将空间气氛提升到极致。为了让餐厅的空间显得更加开阔，设计师还运用了大面积的镜面来装饰。镜面上的雕花纹样细腻、柔美，更为空间增添一缕柔情。

主卧的设计延续了空间的整体风格，沉稳的色调让布艺织物展现出很好的效果。硬包装饰的背景墙、简欧风格的家具、飘逸柔和的织物装饰，都紧扣温馨、浪漫的主题气氛。雕花的木质隔断起到了将卧室与书房空间相

隔的作用，为业主打造出一个利于思考的优雅静谧空间。为了丰富业主的生活，设计师设置了台球室和影音空间，方便业主与好友相聚，利于休闲、娱乐。厨房、卫生间的设计相对简洁、素雅，但依然符合业主的高品质生活，让业主在此享受优雅空间的独特魅力！

The captioned case is Simple-European style. The living room area uses large beige marble with bright colors, giving space a hint of warmth. In furniture styling, European soft lines makes the entire space looks light, romantic and poetic. Restaurant uses calm tone, showing luxurious side.High-quality soft furnishingsenhance the atmosphere of space to the extreme. In order to make the restaurant's space more open, the designer also uses a large mirror as decoration. Carved patterns on the mirror are delicate and soft, adding a ray of tenderness.

The main bedroom design continues the space's overall style; calm tones makes cloth fabric showing good results. Hard pack decorative back wall, simple European-style furniture, elegant fabric decoration, closely link to the warm, romantic topicatmosphere. Carved wooden partitioncuts off the bedroom and study as a role in space apart, creating a quiet and elegant space conducive to think for the owner. In order to enrich the life of the owner, the designer set up a billiard room and audio-visual space, for the convenience of owner and friends' party, leisure and entertainment. Kitchen, bathroom design is relatively simpleand elegant, but still in line with the owner's high quality requirement of life, so that the owner can enjoy the unique charm of this elegant space!

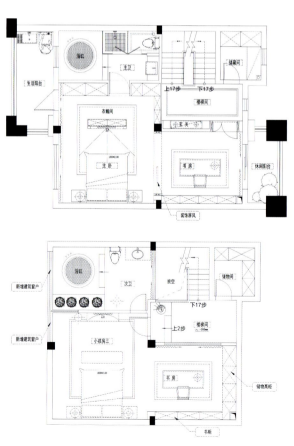

摩登样板间 IV
欧式风尚

顺发康庄某宅

Shunfa Kangzhuang X House

设计单位：杭州周视空间

设计师：周桐

项目面积：120 ㎡

主要材料：不锈钢、艺术壁纸、马赛克

Design Company: Hangzhou Zhoushi Space

Designer: Zhou Tong

Project Area: 120 ㎡

Major Materials: Stainless steel, Art wallpaper, Mosaic

这套房子的每个空间都比较小，因为开发商交付的时候实际上只有两房两卫，而现在改造成有四房两厅两卫。现在餐厅的位置是原来的储藏室。而原先的餐厅在厨房对面，现在改造成儿童房了，在客厅外面的一个区域改成书房了。有很多业主喜欢问能不能把两房改成三房之类，实际上这个东西要看房子的先天条件的，因为改造成一个新的房间首先是要有采光、通风，如果达不到这两个条件，硬隔出的房间一点都不通畅。

本案的主色调是深青色，绿色系中深沉、静谧的色彩。设计师用它营造了一个静谧的世界。整个居室的氛围像是一池湖水，或像是一片让你向往的树林，树荫下是繁星般的小花，在清朗如水的阳光下，舒心地绽放。在客厅中大面积使用深青色，会带来海洋般的神秘、清凉感觉。将它与白色、紫红色搭配，会产生优雅而温馨的感觉。深青色使用在在深邃优雅的客厅空间中，搭配明度和比较高的黄色作为点缀时，强烈的艺术气息瞬间扑面而来而紫色的点缀，让高冷的深青色变得优雅而平易近人。

Each space of this house are relatively small, because actually it only has two bedrooms and two baths, and now it has been transformed into four rooms and two living rooms and two baths. Now the restaurant's place is the storage before. The original restaurant is opposite the kitchen, now it changes into a children room. A zone outside the living room is transformed into a study. There are many owners like to ask whether can change the two rooms into three rooms; in fact, this depends on the natural conditions of the house. Because a successful transformation needs lighting and ventilation, otherwise the partition room will not smooth.

The main color of the captioned case is dark cyan, a deep, quiet color in the green color system. Designer use it to create a quiet world. The whole atmosphere of the room is like a pool of water or a piece of wood, where star-like flowers growing under the green shade, blossoming under the water-like sunshine. In the living room, the large area of dark cyan brings mystery, cool feeling like the ocean. When it matches with white, purplish red it will produce elegant and warm feeling. The dark cyan is used in the elegant living space, with high bright yellow as ornament, then the strong artistic atmosphere makes the cold color elegant and approachable.

摩登样板间 IV
欧式风尚

深圳曦城别墅

Shenzhen Xi City Villa

设计师：龚德成

项目面积：600 ㎡

项目地点：广东深圳

主要材料：影木、金尼斯木、云石、黑金花大理石、壁纸、硬包、玫瑰金不锈钢、马赛克

Designer: Gong Decheng

Project Area: 600 ㎡

Project Location: Shenzhen Guangdong

Major Materials: Shadow wood, Guinness wood, Marble, Black gold flowers marble, Wallpaper, Hard pack, Rose gold stainless steel, Mosaic

MODERN SHOW FLAT IV **EUROPEAN FASHION** 215

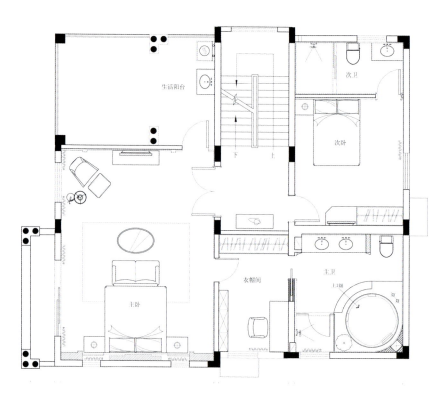

现代社会喧嚣扰杂，人们渴望内心的宁静、恬淡，让都市生活回归自然，心灵融入艺术。对于现代的都市精英而言，表面的形式感并不是生活的第一要素，由内而外散发出来的优雅、舒适，和由此获得的居住幸福感才是此生不渝的追求。

本案的整体设计以简约的设计手法，搭配欧式家具，陈设及艺术品营造自然、现代的装饰美学。

在空间规划上，客厅与会客厅连在一起，采用统一色调的壁纸贯穿两个空间。会客厅墙面用镜面加以装饰，使空间在视觉上显得更开阔，让原本封闭的卫生间生动起来。在餐厅的设计上，设计师以大面积中性色调，搭配局部深咖啡色，形成明暗对比。整体空间的灯光设计以直接照明为主，配上局部光照，让空间散发出质朴、宁静的气质，让身体与心灵同时回归自然。主卧中的蝴蝶兰、花鸟壁纸、中式纹样的台灯，采用了具有东方色彩的元素，更体现出业主的尊贵品位。

The modern society is hustle and bustle. People want inner peace, tranquil, making city life back to nature and the soul into the arts. For the modern urban elites, the superficial formality is not the first element in life. The inner sense of elegant, comfortable, residential happiness are the unswerving pursuit of life.

The overall design of the captioned case uses simple design technique, with European style furniture, furnishings and artwork to create a natural, modern decoration aesthetics.

On spatial planning, living room is linked with the parlor, the unified color wallpaper covers the two spaces. Living room wall is decorated with a mirror, so that the space will be visually more open, and the original enclosed bathroom is vivid now. In the design of the restaurant, the designer uses large neutral color as the tone, matching partial dark brown, forming a chiaroscuro. Lighting design of the overall space bases on direct lighting and partial illumination. The space exudes a rustic and quiet temperament. Therefore the body and mind at the same time return to nature. The butterfly orchid, birds and flowers wallpaper, Chinese patterns table lamp in the main bedroom uses the elements with oriental colors, and also reflects the noble taste of the owner.

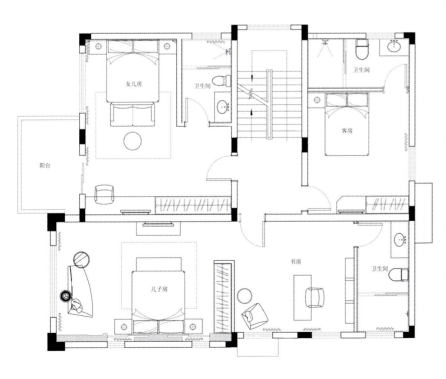

摩登样板间 IV
欧式风尚

古典美学的华丽转身

Splendid Turning of Classical Aesthetics

设计单位：摩登时尚装饰设计

设计师：刘家洋

项目面积：300 m²

主要材料：莎安娜大理石、仿古砖、壁纸、实木等

摄影：李玲玉

Design Company: Modern Fashion Decoration Design

Designer: Liu Jiayang

Project Area: 300 m²

Major Materials: Shana marble, Antique tiles, Wallpaper, Wood, etc.

Photography: Li Lingyu

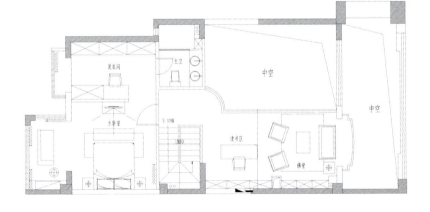

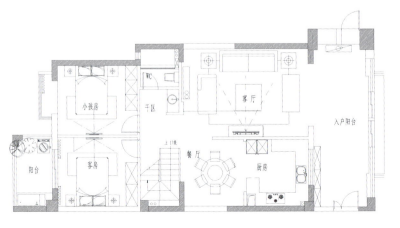

在本案中，色彩丰富的山水绘画、时尚华丽的水晶宫灯、细腻精致的雕花刻金，一派富丽豪华的气象流泻而出。在这里，正上演一场古典主义的华丽转身。设计师以时尚的设计语言来诠释现代人对于古典美感的理解与需求，意欲传递出项目独有的文化气质，呈现出一个低调、奢华的空间的功能性需求。本案方正的平面布局，空间之间流畅地互动，合理的功能分区，都展现出这是设计师精心打造的高端户型。设计师在充分理解原建筑设计空间理念的前提下，对原建筑做了提升、突破，为业主打造了高品质的生活环境。

The captioned case contains colorful landscape painting, gorgeous fashionable crystal palace light, delicate carved exquisite gold. In here, there are putting on a magnificent turning of classicism. Designer uses fashion design language to interpret the modern understanding and needs of classical beauty and demand, intending to deliver an unique cultural temperament, showing a low-key, luxurious functional space. The square layout of the case, smooth interaction between spaces and reasonable functional areas express this is a well-built high-end unit. Designer, with the full understanding of the original architectural design concept, upgrading, breaking a high quality living environment for the owner.

摩登样板间 IV
欧式风尚

天鹅湾某住宅

Swan Bay X House

设计单位：福建国广一叶建筑装饰设计工程有限公司

设计师：林姿宏

方案审定：叶斌

项目面积：100 m²

主要材料：通体砖、大理石、壁纸、茶镜、白色乳胶漆

Design Company: Fujian Guoguang Yiye Architecture Decoration

Design Engineering Co., Ltd.

Designer: Lin Zihong

Design Authorization: Ye Bin

Project Area: 100 m²

Major Materials: Whole body tiles, Marble, Wallpaper, Tea mirror,

White latex paint

本案的设计风格为简欧风格,设计师将打造典雅、自然、高贵的气质,营造浪漫的氛围为设计的目标。生活在繁杂多变的世界里已是烦扰不休,而简单、自然的生活空间却能让人身心舒畅,感到宁静和安逸。借着设计将室内空间解构和重组,便可以满足我们对悠然自得的生活的向往和追求,让我们在纷扰的现实生活中找到平衡,营造出一个令人心驰神往的写意空间。

简欧风格继承了传统欧式风格的装饰特点,吸取了其风格的"形神"特征,在设计上追求空间变化的连续性和形体变化的层次感,室内多采用带有图案的壁纸、地毯、窗帘、床罩、帐幔及古典装饰画,体现出欧式风格华丽的一面。家具门窗多漆为白色,画框的线条部位装饰有金边,在造型设计上既突出了凹凸感,又具有优美的弧线。

The design style of the captioned case is simple-European style. Designer will create an elegant, natural, noble temperament, aiming at a romantic atmosphere as the goal. The complex and changing world sometimes is annoying. However, simple, natural living space is physically and mentally able to make people happy, peaceful and comfortable. By deconstruction and reorganization of the interior space, it becomes a symbol of ideal life of desire and pursuit, letting us find a balance in real trouble life, creating a pleasant relaxation fascinating impressionistic space.

Simple-European style inherits the traditional European style decoration features, especially the body and the spirit. In the design, the case pursue spatial variations continuity and physical changes. The indoor uses patterned wallpaper, carpet, curtains, bedspreads, draperies and classical decorative paintings, reflecting the luxury side of the European-style. The furniture, doors and windows are painted white, and the line of the frame are decorated with golden color, not only emphasizing concave-convex feeling but also has beautiful arc line.

摩登样板间 IV
欧式风尚

奥克斯盛世缔壹城中央样板房

AUX Golden First City Central Show Apartment

设计单位：宁波知行室内设计事务所

设计师：柯其恩、周军华

项目面积：113 m²

项目地点：浙江宁波

摄影：叶建荣

Design Company: Ningbo Zhixing Interior Design Firm

Designers: Ke Qien, Zhou Junhua

Project Area: 113 m²

Project Location: Ningbo Zhejiang

Photography: Ye Jianrong

相遇美好

一个中产阶级的三口之家需要一套怎样的房子？不同爱好的三个人喜欢在怎样的家里生活？缔壹城A户型样板房的设计围绕着这些问题来展开。

这个年轻的家庭里少不了时尚的元素，当然更需要一个温馨的氛围，所以设计师用浅色调搭配金属和皮质材料打造了现代主义的品质空间，在其中融入的新装饰主义元素也令空间的表情显得时尚而生动。

踏入玄关，大理石拼花和钢琴烤漆家具的搭配让时尚、温馨的氛围在餐厅中静默流转，水晶吊灯的优雅，更为此处增添了一丝精致与华美。室内设计通过柔软的布艺巧妙过渡，当人步入浅色调的客厅时，白色皮质的背景墙、金属桌边几与黑色皮革电视柜的时尚搭配完美呈现了一个现代风格的空间。一家人可以舒适地坐在大沙发中，看着电视，享受三口之家的其乐融融。

卧室与书房的设计延续了整体的格调，色彩以浅色和自然色系为主，自然舒适与时尚奢华并存，整体格局简洁、放松，自由的气息流转。徜徉其中，每一步都是美好的，每一天都是快乐的。

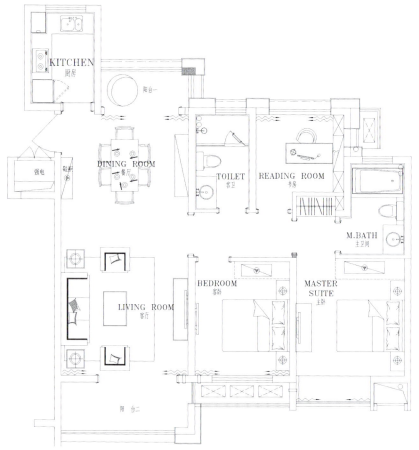

Encounter Beauty

What kind of house does a three people middle-class family needs? What kind of house does the family with different hobbies want to live in? First City A Unit design starts around these problems.

The young family is not lack of fashion elements, and also need a cozy atmosphere, so the designers use light colors with metal and leather materials to build a modernist quality space. The new decoration elements add fashion and vivid sense to the space.

Stepping into the vestibule, the match of marble flower pattern and piano paint furniture make the space full of fashion, warm atmosphere. The elegant chandeliers here adds a touch of refinement and beauty. The interior design uses soft cloth. When people stepping into the light colored living room, the white leather background wall, the metal table and black leather TV cabinet present a modern space together. The whole family can sit comfortably in a large sofa, watching TV and enjoying the leisure time.

The design of the bedroom and study continues the overall style, the main color is light and natural. The overall layout is comfortable, luxury,fashionable, simple, relaxed and free. Wandering in the house, each step is good, each day is happy.

摩登样板间 IV
欧式风尚

情归于家

Charm in Home

设计单位：昶卓设计黄莉工作室
施工单位：怡明施工
软装设计：昶卓软装
项目面积：264 m²

Design Company: Chang Zhuo Design Huang Li Studio
Construction Company: Yee Ming Construction Company
Soft Decoration Company: Chang Zhuo Soft Decoration Company
Project Area: 264 m²

业主在国外生活十多年后，出于对家的牵挂让他们毅然放弃了让人觉得羡慕的工作，回到了祖国，回到了南京。因为在美国多年，美式的休闲与自然让夫妻俩俨然习惯，所以当我们在沟通风格时，看到他们回忆起在美国的生活，如今想起，当时他们的满脸甜蜜到今天依然还能让我感动。

空间的整体色调为温馨的米黄色，客厅的沙发选用了造型简洁的布艺沙发，配合同样低调的木质茶几，展现出业主回归简约生活的心境。空间虽然简约但也不简单，铁艺吊灯、刷白漆的木质顶棚、墙壁上的挂画、茶几上的花艺，无不在说明业主一家的有情调的生活。卧室的设计同样采用浅色调，实木的家具为空间增添了品质感，再搭配温馨的布艺，为业主营造出一个良好的睡眠空间。

看着这样休闲、惬意的家，让人想到了这样的一句话："一切烦琐里才能照见简单，一切世俗中才能照见脱俗，一切喧闹中才能照见沉静。"

After living abroad for more than ten years, because of the care to the home the owners gave up successful career and returned to the motherland, returned to Nanjing. Because they lives many years in the United States, the owners get used to the American leisure and nature. During the communication, they recalled life in the United States, now their sweet face still can make me moved.

Overall tone of the space is warm beige. The living room is selected modeling simple fabric sofa, with wooden coffee table showing the owners' simple mind towards simple life. Space is simple but not dull, iron chandeliers, wooden ceiling whitewashed paint, paintings on the walls and floral coffee table all describe the owners' exotic life. Bedroom design also uses light colors, wood furniture add a sense of quality into the space, the warm cloth creates a good sleeping space for the owners.

Looking such a casual, cozy home, people may think of this sentence: "See the simple from the complicated, see the refined from the secular, see the quiet from the noise."

摩登样板间 IV
欧式风尚

保利塘祁路A户型样板房

Baoli Tangqi Road A Unit Show Apartment

设计单位：上海乐尚装饰设计工程有限公司
设计师：乐尚设计团队
项目面积：82 ㎡
项目地点：上海

Design Company: Shanghai Leshang Decoration Design Engineering Co., Ltd.
Designer: Leshang Design Group
Project Area: 82 m²
Project Location: Shanghai

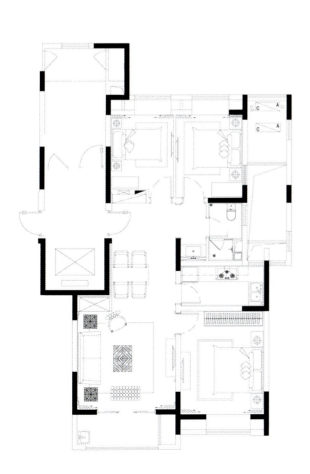

MODERN SHOW FLAT IV EUROPEAN FASHION 253

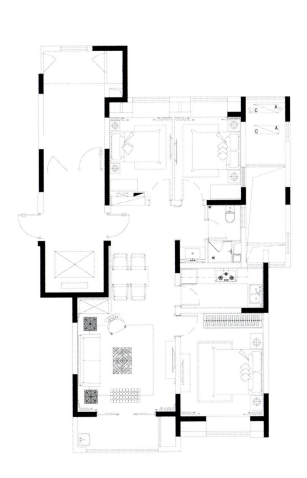

本案设计强调空间的独立性，配饰的选择上也较为稳重大气、唯美典雅，以满足业主的现代生活。空间设计中，设计师使用了中式元素与新古典设计语言相融合，同时加入了现代材质的质感，不拘一格的混搭与随性的设计，让空间真正成为享受生活的家。

In the design of the captioned case, the designer position it as a modern and Neo-classical mixture luxury style according to the characters, ages, consumer groups, social status, preferences, and so on. Neo-classical style is always calmly passing on in high-quality interior design without boundaries or region. Neo-classical design exudes thick form beauty, abandoning complicated decoration, using the most classic and simple way to express the historicity and cultural depth. Besides, Neo-classical style emphasizes coordination between man and man, man and society and improving the building's affinity changing the isolated space into an open one.

The design emphasizes the independence of the space, the choice of accessories is also stable, beautiful and elegant, to meet the modern life of the owner. In the space design, designer uses Chinese elements and Neo-classical design language, while adding a modern material texture. The eclectic mix and casual style make the space enjoyable.

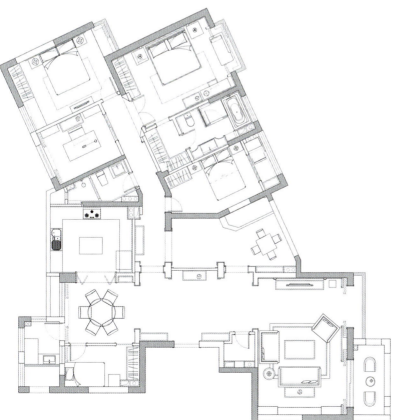

湖北拥有丰富的江河湖泊资源，本案便坐落在最美丽的湖泊——东湖周边。从客厅、花园、主卧向窗外轻轻望去，映入眼帘的便是郁郁葱葱的树林和波光粼粼的湖面，酌上一杯清茶，让人格外惬意。设计 19 ㎡的超大玄关，以香槟色的花鸟题材壁画做背景，将奢华与温馨完美地结合在一起。餐厅与厨房以折叠门相隔，不仅能保持视觉上的通透感，同样能满足对隔绝中餐油烟的要求。在设计厨房时，将其处理为三边带中岛的格局，使用方便、动线流畅。与独立书房相邻的是主人的卧室、衣帽间与卫生间。整体空间的线条通畅且功能统一。主卧墙面采用经典的条纹壁纸配以白色护墙板，与杏色软包浑然一体，典雅气派。精致的家具配上卡其色的窗帘，突显出华丽而典雅的韵致。儿童房采用海蓝色素雅的壁纸，格局合理、功能齐全，使整个空间显得非常温馨、宁静。室内的环境呈现出悠闲、舒畅的生活状态。

Hubei province has rich resources of rivers and lakes. The captioned case locates beside the most beautiful lake——East Lake. Gently looking outside from the living room, the garden, the master bedroom, one can see lush woods and sparkling lake. It will be very comfortable to have a cup of tea. The entrance is 19 m^2 large with champagne flower and bird as background, perfectly combining luxury and warmth together. The restaurant and kitchen are separated by folding door, not only to maintain the transparent visual sense, but also can meet the requirements of isolating Chinese food fumes. In the design of the kitchen, it is treated as a triangular pattern with the counter in the center, making the space easy to use with smooth lines. Beside the independent study is the master's bedroom, cloakroom and bathroom. The smooth lines of the overall space are functional unified. Bedroom's wall uses classical striped wallpaper and white wainscoting, with beige soft pack and elegant style. Fine furniture coupled with khaki curtains, highlighting the gorgeous and elegant charm. Children room uses simple and elegant sea-blue wallpaper, with reasonable structure, complete function, making the entire space warm and quiet. Indoor environment exhibits relaxed and comfortable living conditions.

摩登样板间 IV
欧式风尚

雅致简欧

Elegant Simple European

设计单位：绍兴欧拓者装饰

设计师：潜守政

项目面积：145 m²

主要材料：大理石、实木地板、硬包、黑炭金不锈钢

摄影：右手

Design Company: Shao Xing Tuo Zhe Decoration

Designer: Qian Shouzheng

Project Area: 145 m²

Major Materials: Marble, Wood floors, Hard pack, Black carbon gold stainless steel

Photography: You Shou

本案的设计将传统欧式风格与现代风格相结合，使得整个空间现代简洁而不失奢华，温馨浪漫而不失尊贵典雅。客厅电视背景墙的清新设计，开灯后如同雨后清晨的阳光。大小不一的抱枕，为业主和孩子创造了舒适的环境。深色的茶几上，蓝色花卉、雪茄、红酒与烛台，这些无一不是浪漫生活场景的体现。餐厅的铜质吊灯加上法式浪漫的花卉和餐具，让人们进入宴飨时刻的美妙与温馨。一家人其乐融融围坐在餐桌旁，期待和等候一场全家欢乐宴，这将是多么幸福的生活！

The captioned case is a combination of the traditional European style and the modern style, making the entire space modern, simple and luxurious, romantic, noble and elegant. Living room TV background design is fresh, the lights is like the morning sun. Big and small pillows create a comfortable environment for owner and children. On the dark coffee table are blue flowers, cigars, wine and candelabra, which reflects the romantic scenes of life. Restaurant bronze chandelier plus French romantic flowers and utensils allow people entering into the beautiful and warm banquet moment. The family sits around the dinner table, looking and waiting for a happy family feast. What a wonderful life!

摩登样板间 IV
欧式风尚

Rong Qiao Apartment X House

设计单位：福建国广一叶建筑装饰设计工程有限公司
设计师：叶猛
设计审定：叶斌
项目面积：180 ㎡

Design Company: Fujian Guoguang Yiye Architecture Decoration Design Engineering Co., Ltd.
Designer: Ye Meng
Design Authorization: Ye Bin
Project Area: 180 m²

本案采用现代新古典风格进行设计，平面布局严谨合理，设计师注重公共区域与私密区域的分隔。空间的设计以奢华格调为主，同时也注重布局实用性。整体空间以暖色调为主，客厅中米黄色系的大理石与金箔的顶棚相得益彰，使空间风格趋于统一。卧室则采用紫色系壁纸和床品，为业主营造一个温馨、舒适的睡眠空间。

室内设计通过大量黑色镜面、喷花玻璃、不锈钢等现代材料与欧式线条古典造型相融合，既避免了欧式过于沉重、厚实，又使局部空间尽可能有扩大之感，使整体风格呈现出轻盈、通透的质感和时尚、奢华的气息。

The captioned case uses modern neo-classical style. The layout is rigorous and reasonable. The designer focuses on common areas and private areas' separation. Space's design is luxurious style, while also focusing on practicality. The overall space is in warm tones, the living room's beige color marble and gold ceiling are echoing with each other, so that the space style tends to a unity. The bedroom uses purple wallpaper and bedding to create a warm, comfortable sleeping space for the owner.

The interior design uses a large number of black mirror, spray glass, stainless steel, other modern materials and classical European lines to avoid the European style's heavy and thick sense and make the local space visually expands. Therefore the overall style will be light, transparent, stylish and luxurious.

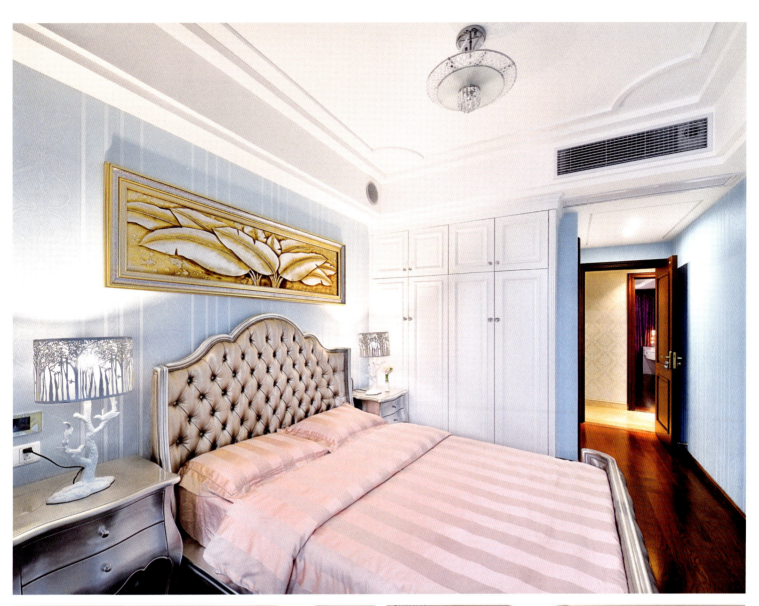

摩登样板间 IV
欧式风尚

中海万锦江城某样板间

Zhonghai Wanjin River City X Show Apartment

设计单位：广州市韦格斯杨设计有限公司

项目地点：湖北武汉

主要材料：壁纸、木地板、软包、灰镜

Design Company: Design Company: GrandGhostCanyon Designers Associates Ltd.

Project Location: Wuhan Hubei

Major Materials: Wallpaper, Wood floors, Soft bag, Gray mirror

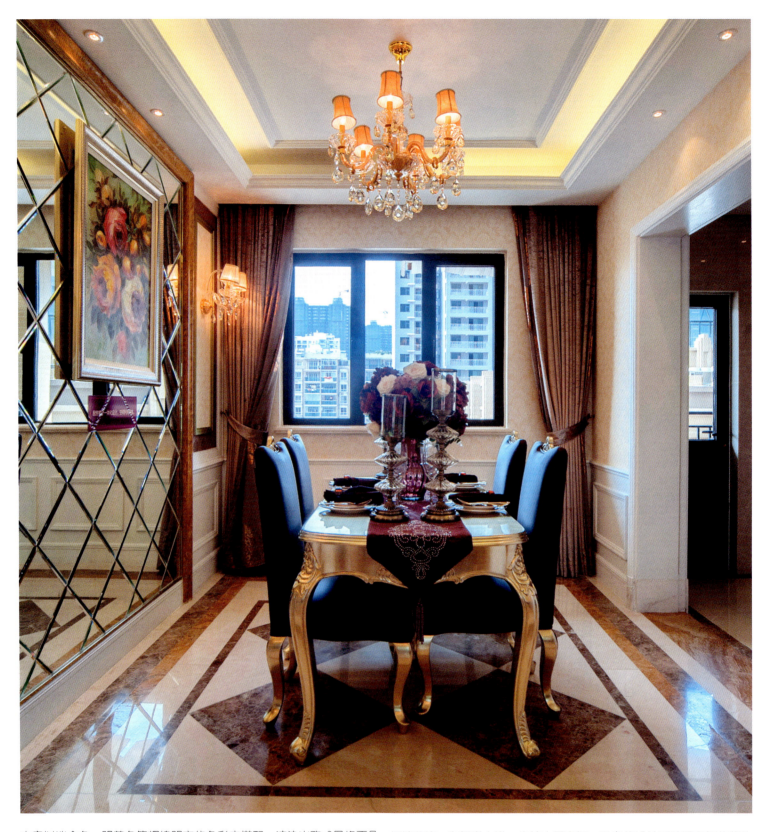

本案以淡金色、明黄色等辉煌明亮的色彩来搭配，渲染出欧式风格不凡的气度。镶花刻金的墙饰与纹理复杂的装饰品，均以一种雍容、华贵的姿态，传递着业主高雅的审美情趣和极富贵族文化底蕴的生活态度。本案的客厅与餐厅空间相互贯通，动线自如、视野开阔。餐厅空间更是以银镜装饰，在视觉上进一步扩大了空间。卧室的设计延续了空间的整体风格，却没有拘泥于此，床头硬包的搭配使空间更显奢华，飘窗的设计也更显情调。本案的设计是为业主量身打造的奢华空间，不仅带来了欧式优雅尊贵的生活感受，更是一圆业主的宫廷贵族梦。

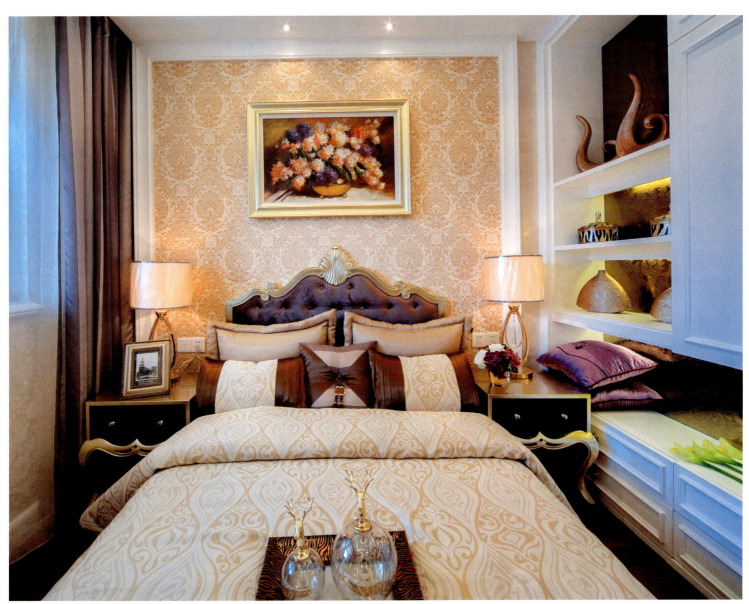

The case mainly uses pale gold, bright yellow and other bright colors, rendering European-style extraordinary tolerance. Parquet carved gold wall hangings and decorations with complex texture are conveying the owner's elegant, aesthetic, and noble attitude towards life through a kind of grace, luxurious gesture. The case's living room and dining room are connected with each other with smooth generatrix and wide vision. The restaurant space is decorated with silver mirror, visually expand the space. Bedroom design continues the overall space style, but not rigidly adhere to this. The bed hard pack makes the room more luxurious, and the windows more exotic. The case design is a luxury space tailored for the owner, not only brings European elegant noble life experience, but also makes the owner's aristocracy dream come true.

摩登样板间 IV
欧式风尚

眉山凯旋国际广场样板间

Meishan Triumphant International Square Show Flat

设计单位：深圳创域设计有限公司、殷艳明设计顾问有限公司
设计师：尹艳明
项目面积：95 ㎡
主要材料：乳胶漆、壁纸、木地板、地砖、石材
开发商：四川领地房地产集团股份有限公司

Design Company: Shenzhen Chuangyu Design Co., Ltd. /
Yin Yanming Design Consultant Co., Ltd.
Designer: Yin Yanming
Project Area: 95 m²
Major Materials: Latex Paint, Wallpaper, Flooring, Tiles, Stone
Developer: Sichuan Lingdi Real Estate Group Corporation

本案灵感来自迷人的地中海风格，以现代审美打造古典韵味，设计师用简单的设计理念捕捉光线、大胆而自由地运用色彩，装饰元素的样式、取材多来自大自然，居住环境中充盈着蓝天、碧海的安宁、静谧。

进入玄关，正对面是装饰鞋柜，丰富的层板设计满足了家庭的需求。客厅中央美丽的吊式水晶灯绽放出耀眼夺目的光芒，显现出华丽的气质。茶几下的蓝白相间地毯在水晶灯的照射下非常美丽。空间中蓝色与白色无处不在，好像薄纱一般轻柔，让人感到自由自在。背景墙上的陶瓷挂件让人想起碧海、蓝天连成一片的情景，在这仿佛听见了海风在歌唱，带来一朵浪花、一缕海风，想像在风中嬉闹，在浪中沉醉。

卧室的设计在色调上以白色、木色、蓝色为主，再通过中间色系的过渡配以咖啡色点缀，使整个空间在极具视觉冲击力的同时又不失和谐，彰显出温馨而舒适的一面。将外在的自然环境融入室内环境当中，使得整个空间更加亲近自然，从而达到人与自然的完美结合。在空间的设计上不仅考虑了休闲功能，同时也增加了空间使用与储藏功能，达到功能与形式的有机结合。

在整体设计上，到处充满着温馨，让人心情舒畅。简单、质朴的墙面肌理渲染着地中海的情怀，大自然的色彩就在这样的背景上展开。装饰元素相互呼应和延续令整个空间在变化中拥有统一的特质，让温情无处不在。

The inspiration comes from charming Mediterranean style, creating classical charm with a modern aesthetic. The designer uses simple design concept to capture the light, boldly using colors, decorative elements from the nature. The living environment is full of tranquility and quietness of the blue sky and ocean.

Stepping into the entrance, one can see the decorative shoe cabinet right opposite the entrance. The rich laminates design meets the need of families. The beautiful hanging crystal lamp at the central living room shines a dazzling light, revealing a gorgeous temperament. Blue and white rug under the coffee table seems very beautiful under the illumination of the crystal lamp. The space is full of blue and white, like general soft chiffon, making people feel free. The ceramic ornaments on the background wall reminiscent people of sea, sky together into one scene, in which one almost hear the singing of the sea, the spray, the wisp of wind. The imagine frolics in the wind, basks in the waves.

Bedroom design bases on white, wood, blue color, through the transition of the middle color system, the designer matches the space with coffee color. The entire space is very visually impact, but also harmonious at the same time. The outside of the natural environment has been add into the indoor environment, making the space more close to nature, so as to achieve the perfect combination of man and nature. The space design considers not only the recreational functions, but also increases the use of space and storage functions, reaching the combination of function and form.

The overall design is full of warmth, making people feel comfortable. Simple, rustic wall texture renders Mediterranean feeling, the color of nature had begun in this context. Decorative elements echo each other and continue to make the entire space with a unified and warm nature in the change.

摩登样板间 IV
欧式风尚

红磡海逸豪园

Hunghom Harbour Plaza Garden

设计单位：擎风设计
设计师：林子康
项目面积：148.6 ㎡
项目地点：香港
主要材料：定制木器家具、玻璃、壁纸

Design Company: Qingfeng Design
Designer: Max Lam
Project Area: 148.6 m²
Project Location: Hong Kong
Main Material: custom wood furniture, glass, wallpaper

客、餐厅以英式风格和现代风格为主，以大理石配搭清镜，彰显大气。材料运用的一致性将两厅空间相互融合，再配以两盏雅致、独特的水晶灯，让人感受到优雅的气息。两间卧室以简约为主题，浅色的搭配给房间增添了不少生气。楼梯台阶既是通道，也是隐藏的储物空间。主卧的视线焦点在特色床头上，绘有不规则图案的扪皮由床头直至顶棚，有型又大方。主卫和客卫均使用大理石作为墙面、地面的材料。为避免宽阔的空间过于单调，设计师对于地面及镜柜的设计花了不少心思。而厨房的设计，设计师大胆以深色木纹作为橱柜的饰面，既独特又不失舒适、和谐。

Guest, dining-room is given priority to with English style and modern style, with the marble mirror, highlight the atmosphere. Materials using the consistency of the two hall space mutual confluence, match again with two refined, unique crystal lamp, let a person feel elegant breath. Two bedrooms with contracted as the theme, of light color collocation adds a lot of the room angrily. Stair steps is channel and hidden storage space. Advocate lie the line of sight of the focus on the characteristics of the head of a bed, with irregular pattern of the ammonites skin by until the ceiling of the head of a bed, have a type and generous. Advocate defend and the guest is defended use marble as the material of metope, ground. In order to avoid the broad space too drab, stylist for the ground and the design of the lens ark spent a lot of mind. And the design of the kitchen, stylist is bold with brunet wood as veneer of ambry, both unique and comfortable and harmonious.

摩登样板间 IV
欧式风尚

漫步人生

Stroll On The Way of Life

设计师：王勇

项目面积：120 m²

主要材料：中花白大理石、壁纸、水曲柳、全抛仿古砖、软包、灰镜

Designer: Wang Yong

Project Area: 120 m²

Major Materials: Zhonghua white marble, Wallpaper, Ash, Whole cast antique tiles, Soft bag, Grey mirror

本案为120㎡的简欧平层空间，设计师将欧式古典元素以现代手法诠释，抛弃多余繁杂的点缀，在整体格局的基础上用跳跃性细节来做变化，整个空间洋溢着典雅的气质、浪漫的情调。空间的设计仿若宫廷舞步的细致与浪漫，具备不同的气质与姿态，用跳跃的变化渲染着整体的氛围。

在空间布局上，设计师合理安排功能布局，强调空间整体协调关系，在保持足够的储藏收纳空间的同时，让其他功能的区域空间尽可能的放大。在软装上，以米白色系定位基调，雅致的家具、精美的饰品为搭配，为崇尚高品质生活的业主打造出自然、素雅的简欧居所。各种饰品和艺术收藏将装饰性和实用性完美地结合在一起，通过吊灯、软包、家具、台灯、雕塑及植物，不仅为家庭营造了艺术氛围，而且表达了一种优雅的生活态度。

空间每一个画面，每一处细节，简而不失其华，约而不失其涩。

The case locates for 120 m², the style is simple European, the designer add European classical elements in the modern techniques interpretation, abandoning unnecessary complicated embellishment, using jumping detail on the overall pattern, so that the entire space is filled with elegant and romantic atmosphere. The space design is like palace steps—full of meticulous and romantic sense, with a jump of changes and different temperament and attitude.

For the space layout, designer reasonably arranges the function layout, emphasizing the overall coordination of the space, while maintaining sufficient storage space and enlarging other functional regional space. As for the soft decoration, the designer uses white color as the tone, then matches it with elegant furniture, fine jewelry and so on, creating high-quality, natural and elegant European style home for the owner. A variety of jewelry and art collections are decoratively, practically and perfectly combined. By chandeliers, soft bag, furniture, lamps, sculptures and plants, the designer not only creates an artistic atmosphere for the family, but also expresses an elegant attitude towards life.

Every picture in the space and every detail are simple but not dull.

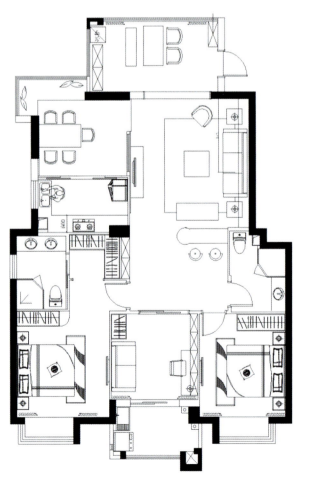

和昌森林湖 G2-6 户型样板房

Hechang Forest Lake G2-6 Unit Show Apartment

设计单位：上海轶骐设计工程有限公司
参与设计：黄美绮、王宗仁
项目面积：87.43 ㎡
主要材料：仿古砖、实木地板、砂岩漆、壁画、涂料、铁艺等

Design Company: Shanghai MAXIMA Design Engineering Co., Ltd.
Participatory Designers: Huang Meiqi, Wang Zongren
Project Area: 87.43 m²
Major Materials: Antique brick, Wood floors, Sandstone paint, Murals, Paint, Iron, etc.

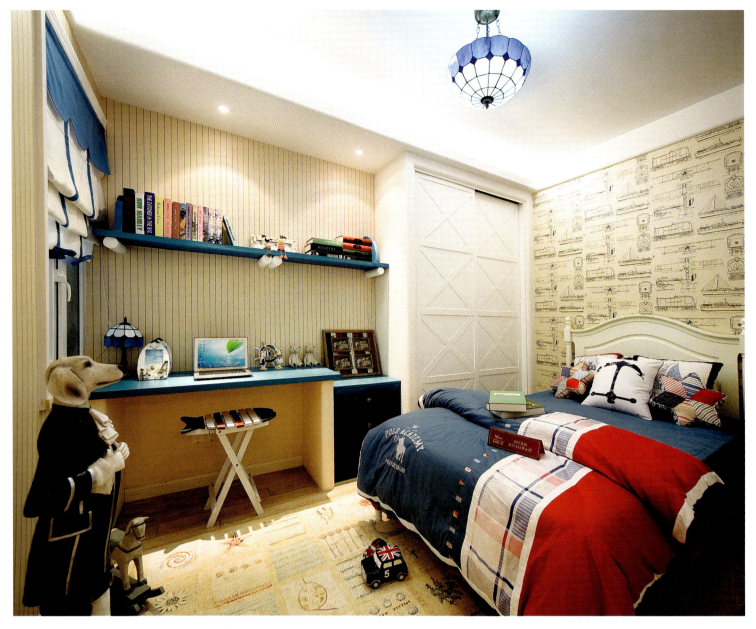

本案位于低密度小区，周边环境优美。原户型为两房两厅户型，将业主定位为一对年轻的夫妇，有一个4岁的儿子。室内设计风格定位为地中海风格。都市中繁重的工作压力，过快的生活节奏常常让人迷失，变得盲目。而当地中海风格来袭时，处处都透露着惬意与舒适的生活趣味。

本案的设计像是将地中海沿岸休闲的亲水生活带入了空间，让人领略到的不仅是悦目，更是动心的宁静。步入门厅的那一刹那，你能感觉到仿佛有一股迎面吹来的海风，带着海水的气息。

设计师在布局结构上颇具心思，在原有结构中，厨房是独立的，卫生间门直对餐厅，使得餐厅空间比较凌乱，厨房的空间也比较局促。因此，设计师将原有的厨房改成开放式厨房，把卫生间门改到与厨房同一方向开，并把餐厅墙面做了拱形处理，把地中海风情壁画带入其中。菱形花格的屏风将餐厅空间和厨房空间进行了视觉上的分隔，大大的增强了空间感和视觉效果。

The captioned case is located in a low density area with beautiful surroundings. Original unit contains two bedrooms and two living rooms for a young couple and a 4-year-old son. Interior design style is Mediterranean. The city heavy work pressure, fast pace of life often make people lost and blind. Mediterranean style creates pleasant and comfortable life.

Design of the case is like bringing the Mediterranean coast leisure life into a space.People can feel pleasant and quiet there. When stepping into the hall,one can feel an ocean breeze blowing ahead with sea flavor.

Designer considers a lot in the layout structure.In the original structure, the kitchen is separate, the bathroom door straightlyfaces the restaurant, making the restaurant space messy.The kitchen space is relatively narrow. Thus, the designer changes the original kitchen into an open kitchen, the bathroom door and the kitchen door are open in the same direction.The restaurant walls is arch, the designer brings Mediterranean-style into it. Diamond plaid screen separates the dining space and kitchen space visually, greatly enhances the sense of space and visual effects.

Hechang Forest Lake G2-5 Unit Show Apartment

设计单位：上海轶骐设计工程有限公司

设计师：黄美绮、王宗仁、肖映东、樊晶

项目面积：82.23 ㎡

主要材料：地砖、白蜡木地板、白色混水漆、皮革、银镜等

Design Company: Shanghai MAXIMA Design Engineering Co., Ltd.

Designers: Huang Meiqi, Wang Zongren,

Xiao Yingdong, Fan Jing

Project Area: 82.23 m^2

Major Materials: Brick, Ash wood floors, White mixing water paint,

Leather, Silver mirror, etc.

本案为湖景小高层，原户型为两房两厅，由于是小户型，空间较为局促，设计时将阳台纳入客厅；减少了厨房与卫生间面积，使餐厅归为一个视觉更大的椭圆形；在餐厅加了一个百叶窗以营造出更好的氛围。

空间整体采用米色、白色等浅色调，灯光主要采用泛光形式，局部采用聚光，使空间在视觉上更为宽敞。客厅的顶棚设计将原有阳台的梁做了处理，使得客厅的空间延展性更强。餐厅的装饰造型随平面的形式，加强了区域感。在局部采用了银镜，营造出奇妙的空间感觉。

The captioned case is lakeview high-rise building, the originalunitcontains two houses and two rooms. Since the house is small size, the space is relatively narrow.The balcony is design incorporated into the living room; reducing the kitchen and bathroom area.The restaurant is classified into as a largervision.In the restaurant the designer plus a louver to create a better atmosphere.

Space overall uses beige, white and other light colors.Lighting mainly uses flood and partially uses spotlight. The ceilingdesign of the living room changes the original balcony beam, making the living space ductility stronger. The restaurant decoration enhancing the regional sense. The space locally uses the silver mirror to create a wonderful spacefeeling.

摩登样板间 IV
欧式风尚

恬静心居

Quite Heart Space

设计单位：李建林室内设计工作室
施工单位：曾俊文施工团队、阿炳
设计师：李建林
摄影：李玲玉

Design Company: Li Jianlin Interior Design Studio
Construction Company: Zeng Junwen Construction Group, Bing
Designer: Li Jianlin
Photography: Li Lingyu

MODERN SHOW FLAT IV **EUROPEAN FASHION** 309

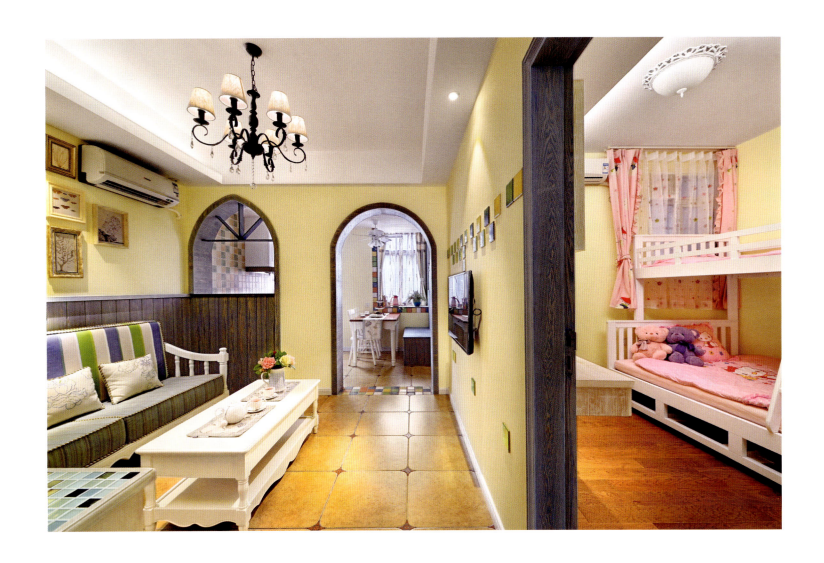

家，对于很多人来说是港湾，也是梦安之处。本案的业主偶然间看过设计师的作品，见面后寥寥数语就确定了由他来设计。业主的定位是带点田园风格的混搭，这也与业主自身恬静、随和的气场相一致。

房子是20世纪90年代的单位房，采用了混砖结构建造的。最大的问题是入户过道窄、无独立的餐厅、厨房也偏小。由于原本结构不方便改动，因此在格局上也不能有大的变动。设计师在空间布局上，把入户过道拓宽，增加了鞋柜和冰箱位置。将厨房与餐厅相结合，厨房设计为开敞式，餐厅设计为卡座式，使空间美观又实用。但设计不能做到十全十美，空间中唯一小小遗憾是卫生间无独立的淋浴房。这皆在于"取舍"两字，设计如此，人生亦是如此。对于一套成型的作品来说，设计只是一部分，业主与设计师之间的配合才是关键。

Home is a haven for many people, and also the dream of safe place. The owner happened to read the designers'case.Through a few words on meeting the designer determine to handle the case. The owner's favorisMix and Match with a little rustic style, which is also the owner's own quiet, easy-going aura.

The house is a 1990s unit, using a mix of brick construction. The biggest problem is the narrow aisles, no separate dining room, and the over-small kitchen.Since it is inconvenient to change the original structure, so the structure can not be majorly changed. Designer broadens the aislesand increase the shoe and refrigerator location. The kitchen and dining room are combined together, the kitchen is open, restaurant is deck typemaking the space beautiful and practical. The design can not be perfect, the only regret in space is the bathroom which has no separate shower room. Design and life are all lies in the choice of accept and reject. For a molding works, the design is only a part, the key lies in the fit between the owners and designers.

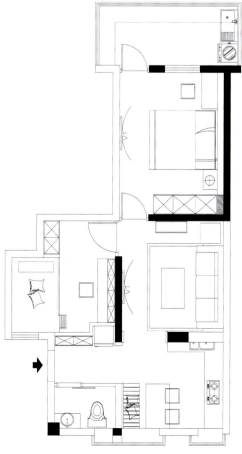

摩登样板间 IV
欧式风尚

清韵昂然

Full of Charm

设计单位：昶卓设计
施工单位：怡明施工
软装设计：昶卓软装
项目面积：220 m²

Design Company: Chang Zhuo Design
Construction Company: Yiming Construction
Soft Decoration Design: Chang Zhuo Soft Decoration
Project Area: 220 m²

设计师曾写过这样一句话："对于每一个设计师来讲，作品就像是自己的孩子一样需要培养、呵护，才能让自己的作品不断的成长，变得更优秀。"而业主也深有感触，一直在忙碌其实就是为了在装修结束时，让家人能有一个满意、舒心的家。

在这个家里，盎然的花韵拂拭了心境沉积的尘埃，滋润着干涸已久的心灵田野。抹去了急功近利的浮躁，给我们流浪漂泊的心灵注入馨香。这就是女主人最想拥有的家的模样。

首先进入眼帘的就是门厅区域，左手的鞋柜故意做得相对较低，就是为了方便于进门换鞋用，而旁边的高柜是便于放衣服，使美观与实用功能相结合。

在客厅中，电视背景墙旁的玻璃隔断上的花形，与电视背景墙的壁纸是完全一致的，这都是设计师精心设计的。沙发背后内嵌的马赛克，让空间变得更有层次。

考虑到餐厅的储藏需求，设计师在餐桌旁做了一组酒柜，既实用又显得大气。餐厅顶棚运用了不锈钢，让降低的顶棚不会显得压抑。

Designer has wrote a sentence: "For every designer, the work is like their own children who needs training and care before continue to grow and become better." The owner also emotionally agree, since the busy work is for their families can have a satisfying, comfortable home.

In this house, flowers wipe the dust deposited in mind, and nourish the soul on the dry field, erasing the quick success of the impetuous, giving our vagrant mind fragrance. This is the hostess most like to have as a home.

Firstly the foyer area enters into the eye, the left shoe cabinet intentionally is made relatively low, it is for the convenience of changing shoes, while the high cabinets are design for clothes, making beautiful and practical combined together.

In the living room, flower-shaped glass partition beside the TV background wall is in consistent with the wallpaper of the TV background, which are well designed by the designer. Mosaics embeds behind the sofa, so that the space becomes more structured.

Considering the restaurant's storage needs, designers make a set of wine cabinet beside the table, both practical and warm. The restaurant roof uses stainless steel, so that the reducing ceiling will not seems depressed.

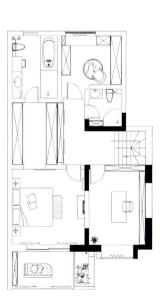

本书在编写过程中，得到各位参编老师的倾力协助，特表示感谢，以下为参编人员名单（排名不分先后）：

唐震江	郭　颖	吕荣娇	欧阳云	张　淼	王丽娜	王　寅	夏辉磷	华　华	贾　蕊	廖四清	
葛晓迎	高　巍	张莹莹	张　明	张　浩	梁敏健	邓　鑫	刘升山	刘　斌	许友荣	郫春园	
史樊英	史樊兵	吕　源	吕荣娇	吕冬英	张海龙	段栋梁	孙朗朗	张　艳	李美荣	陈靖远	
宋献华	吴源华	朱臣高	鲍　敏	刘勤龙	付均云	胡荣平					

图书在版编目(CIP)数据

摩登样板间. 第4辑. 欧式风尚 / ID Book图书工作室编 —武汉：华中科技大学出版社，2015.8
ISBN 978-7-5680-0662-0

Ⅰ. ①摩… Ⅱ. ①I… Ⅲ. ①住宅－室内装饰设计－图集 Ⅳ. ①TU241-64

中国版本图书馆CIP数据核字(2015)第039043号

摩登样板间Ⅳ 欧式风尚　　　　　　　　　　　　　　　　　　　　　　　　　ID Book图书工作室　编

出版发行：华中科技大学出版社（中国·武汉）
地　　址：武汉市武昌珞喻路1037号（邮编：430074）
出 版 人：阮海洪

责任编辑：曾　晟　　　　　　　　　　　　　　　　　　　　　　　　　　　　责任监印：秦　英
责任校对：赵维国　　　　　　　　　　　　　　　　　　　　　　　　　　　　装帧设计：张　艳

印　　刷：深圳当纳利印刷有限公司
开　　本：965 mm×1270 mm　1/16
印　　张：20
字　　数：288千字
版　　次：2015年8月第1版第1次印刷
定　　价：348.00元(USD 69.99)

投稿热线：(010)64155588-8000
本书若有印装质量问题，请向出版社营销中心调换
全国免费服务热线：400-6679-118　竭诚为您服务
版权所有　侵权必究